THE LEGEND

OF

COLTON H. BRYANT

THE LEGEND

OF

COLTON H. BRYANT

Alexandra Fuller

SIMON &
SCHUSTER

London · New York · Sydney · Toronto

A CBS COMPANY

First published in Great Britain in 2008
by Simon & Schuster UK Ltd
A CBS COMPANY

1 3 5 7 9 10 8 6 4 2

Simon & Schuster UK Ltd
Africa House
64–78 Kingsway
London WC2B 6AH

www.simonsays.co.uk

Simon & Schuster Australia
Sydney

Excerpt from "Feed Jake" by Danny Bear Mayo. © 1990 Sony/ATV
Music Publishing LLC. All rights administered by Sony/ATV Music
Publishing LLC, 8 Music Square West, Nashville, Tennessee 37203.
All rights reserved. Used by permission.

A CIP catalogue for this book is available
from the British Library.

ISBN: 978-1-84737-275-8 (Hardback)
ISBN: 978-1-84737-354-0 (Trade paperback)

Designed by Amanda Dewey
Printed and bound in the UK by
CPI Mackays, Chatham ME5 8TD

For Dakota and Nathanial
Because of C.H.B.
From Justice to Forgiveness

Feed Jake

I'm standing at the crossroads in life, and I don't know where to go.
You know you've got my heart babe, but my music's got my soul.
Let me play it one more time, I'll tell the truth and make it rhyme,
And hope they understand me.

Now I lay me down to sleep, I pray the Lord my soul to keep.
If I die before I wake, feed Jake, he's been a good dog,
My best friend right through it all, if I die before I wake,
Feed Jake.

Now Broadway's like a sewer, bums and hookers everywhere.
Winos passed out on the sidewalk, doesn't anybody care?
Some say he's worthless, just let him be.
But I for one would have to disagree.
And so would his mama.

Now I lay me down to sleep, I pray the Lord my soul to keep.
If I die before I wake, feed Jake, he's been a good dog,
My best friend right through it all, if I die before I wake,
Feed Jake.

If you get an ear pierced, some will call you gay.
But if you drive a pickup, they'll say "No, he must be straight."
What we are and what we ain't, what we can and what we can't,
Does it really matter?

Now I lay me down to sleep, I pray the Lord my soul to keep.
If I die before I wake, feed Jake, he's been a good dog,
My best friend right through it all, if I die before I wake,
Feed Jake.

If I die before I wake, feed Jake.

CONTENTS

———⊷⊶⊷———

CONTENTS

CONTENTS

CONTENTS

CAST OF CHARACTERS

Colton H. Bryant—Wyoming boy
Melissa—Colton's wife
Nathanial—Melissa and Colton's son
Dakota Justus—Melissa and Colton's son
William Justus Bryant (Bill)—Colton's father
Kaylee Bryant—Colton's mother
Preston—Colton's older brother
Mandi—Preston's wife
Tabby—Colton's older sister
Tony—Tabby's husband
Merinda—Colton's younger sister
Shad—Merinda's boyfriend
Jake—Colton's best friend
Tonya—Jake's wife
Cocoa—Colton's horse

THE LEGEND

OF

COLTON H. BRYANT

PART ONE

I

A WESTERN

This is the story of Colton H. Bryant and of the land that grew him. And since this is Wyoming, this story is a Western with a full cast of gun-toting boy heroes from the outskirts of town and city-shoddy villains from head office. There is a runaway mustang and crafty broncos. There are men worn as driftwood and salted women and broken-hearted oil rigs. And in this story, the wind is more or less incessant and the light is distilled to its final brightness because of all the hundreds of miles it must cross to hit the great high plains. And the great high plains themselves, dry as the grave in these drought years, give more of an impression of open sea than of anything you could dig a spade into. A beautiful drowning dryness of oil.

But like all Westerns, this story is a tragedy before it even starts because there was never a way for anyone to win against all the odds out here. There's no denying that like the high seas, the high plains of Wyoming make for a hungry place, meanly guarding life, carelessly taking it back. No crosses count. Ground blizzards in the winter and dust storms and wildfire smoke in the summer, everything turning into a sameness of grey so that between the edge of the road and the rest of Wyoming, between earth and sky—there are times a person has no way to tell the difference.

And in this story . . . Well, someone is *always* dying to make room for the next wave of people who are trying to find a way to get rich on all this impression of endlessness out here. Therefore, in this story there is death. Which is nothing new or old in Wyoming and eventually we too—the storytellers and storytold—will go the way of the Indians, the buffalo, the cowboys, and the oil men. We too will make room for someone or something new. An unpeopled silence, perhaps.

2

COLTON AND THE
KMART COWBOYS

Evanston, Wyoming

———◆———

Here is Colton H. Bryant at eight years old pedaling so pitiful fast through the streets of Evanston, Wyoming, that his legs look like eggbeaters. He has white-blond hair and he's tanned the color of stained pine and even at this speed—even at a distance—you can see the color of those eyes. They're such a stunning shade of blue that they register as an absence, like a washed, empty sky. But right now there are tears flooding from those eyes and streaking down Colton's cheeks as he leaps curbs and ducks into side streets, his heart going like a piston, like it would keep beating even if it were torn out of his chest and left alone in all these wide, high plains.

"Retard!"

Colton rides with more sense of panic than direction. He is sawing back and forth across town, past the Dairy Queen and the Taco Bell, up Sage Street, down Summit, over the patch of sun-burned grass behind the old railway station. But everywhere he goes is cluttered with its quota of bored little Kmart cowboys, so called because maybe they docked a lamb's tail for 4-H once a year and maybe they'll grow up to wreck a groin muscle riding the odd bull at a small-time rodeo, but these boys aren't cowboys now and they won't grow up to be cowboys either. You have to have a heart for that, and these boys are bred heartless and made more heartless by the poverty of their imaginations.

"Retard!" they call him because Colton's in special ed and that's on account of the way his brain works, like a saddle bronc, fired up for eight seconds maximum and then bolting around the rails looking for a way out of the arena. Even on Ritalin, Colton has a way of tearing out of the chute, firing with all hooves at once. Colton doesn't have the gear between flat out and stopped. He doesn't have speed perception—the way other people feel alarmed when they're going too fast, Colton feels alarmed when he isn't moving fast enough.

Colton puts his hand up in class one day.

"Yes, Colton?" says his teacher. "You have a question?"

"No, ma'am," says Colton. "It's more of a suggestion."

"Yes?"

"Well, ma'am, I was just wondering if you could talk twice as fast and then we'll get 'er done twice as quick and then we can get out of here in half the time."

And all the other kids start laughing and Colton looks around. "What? What'd I say?"

And the teacher says, "Colton H. Bryant, would you take a deep breath and count to ten and hold your horses?"

• • •

Colton keeps pedaling.

"You're a retard!" comes the shout from a lookout post near the laundry where Colton's dad takes his greasers when he comes back every other week off the rigs so he doesn't clog up the machine at home with all the mud and oil from work. And for a moment Colton pictures Bill at the door of the laundry, all immovable in his broad black cowboy hat, and a lump hurts the front of Colton's throat, but then the light shifts and the image of Bill shifts too, taking with it all that rough Wyoming justice.

"Retard!"

Colton takes one hand off the handlebars long enough to wipe his nose. Evanston is getting kind of blurry. He starts to weave his way recklessly in and out of the streetlights like they were barrels to clear, leaping the curb right in front of cars. Horns blow and in an hour Kaylee will get another phone call from a neighbor telling her that Colton was seen riding recklessly through town. But Colton doesn't care.

"What a freakin' retard!" is what he hears.

Colton's chest fills up with something—he's not sure what it is—because he isn't angry and he's beyond feeling sad and he's too young to know what forgiveness feels like. Then suddenly, "It's okay," he shouts over his shoulder, his voice all high and broken with tears. "Mind over matter. I don't mind so it don't matter." Colton heard that somewhere once, on television maybe, and he likes the magical ring of it. It's like an invisible cloak, the power of not minding anything. Colton's legs whip around and around, "Mind-over-matter; mind-over-matter; mind-over-matter" is the rhythm.

He soars below the underpass and up into the part of town where the hooty-tooty-almighty folk live. His pockets are full of knuckle-sized rocks painted by Merinda and Tabby. Colton is supposed to be selling them for a quarter each, fifty cents if the folk look rich enough. A buck if they seem really stinkin', rollin', filthy. But now his sisters are gonna give him a hard time for not selling rocks and his brother, Preston, is probably just gonna plain give him a beating with no good excuse. "I'm dead," thinks Colton and when he thinks about being dead that makes him think of cowboys and when he thinks of cowboys his mind skips straight to mustangs, which is part of the beauty of Colton's mind. It hardly ever sticks around in one place long enough to get too sad or stay too mad.

"Whee-haw," says Colton, letting his bike have her head. "The Injuns are coming! The Injuns are coming!" he yells, scaring himself for real a little bit at the thought of all those bloodthirsty braves on his tail. And now, under his very seat, the bike transforms itself into a mustang, barely broke, stretching her head across the prairie faster than any other horse in the whole wide West and no one can catch Colton now, not Injuns, not Kmart cowboys, not Merinda and Tabby, not Preston, not anybody. "Come on girl," Colton tells his bike, "let's get outta here."

3

PRESTON AND COLTON, HUNTING

So when they were young—Preston was five years older than Colton—Preston could do any amount of damage to Colton and Colton just smiled right on through it. For example, Preston threw Colton down the stairs with a cushion tied to his waist and Colton laughed all the way down and came running back up for more. Another time, a couple of years later, Preston roped Colton and dragged him all over the yard until the seat of his pants was worn clear through to his boxers and all the time Colton giggling, "He-he-he!" And he was still laughing after Kaylee came home and gave them both a whupping for ruining new clothes. "I got two tons of trouble," she used to say. "Prest-ton and Colt-ton." And every year after Christmas lunch, Bill would pull his boys with ropes behind the pickup on an icy road so that they could ski on the heels of their cowboy boots, and that was always funny, although Preston was trying to trip up his brother. And even when a fight would get out of control and result in a broken nose, Colton's reaction was always the same. "Mind over matter," he said, his eyes swollen shut and his nose all wrapped in gauze.

But then Preston left home and started working out on the rigs and he missed his kid brother and he started to feel bad about all the beatings he'd given him, so that fall, a few months after Colton's fourteenth birthday, Preston announced he was going to take his brother hunting—and by hunting he meant deer and elk, which is entirely different from shooting jackrabbits and geese. And of course Colton said, "Holy cow," and was packed and ready to go for about three weeks ahead of time. He cleaned his gun so hard it's a miracle he didn't rub the metal right off the barrel, he sharpened his hunting knife until it sang, he wore his orange wool hat all day and night, and he lay awake visualizing himself in front of a nice five-pointer, calmly taking it down with one shot.

The night before the hunt, the men stayed up around the table after supper—Bill, Preston, Grandpa, and Colton—and they told Colton their hunting stories and Colton got so excited and nervous he just about hopped out of his own skin. There isn't a decent way to translate to most people the importance of a boy's first hunt, because it's all life's lessons rolled into one day: endurance and tolerance; having the heart to kill and the soul to feel awed by another creature's death; containing yourself and learning to be still; silence and companionship.

"We'll go get you a nice little buck," promised Preston.

"Elk," said Colton. "I want to hunt an elk."

Preston spat. "Let's begin at the beginning, kid."

Bill laughed. "That's right son, one step at a time."

But then Grandpa interjected about the time a fella he knew was out hunting a deer only to find a lion had been trailing him the entire day, and that reminded Preston about the bears up there on the pass, and so it went until Kaylee came into the kitchen and said, "Now you men have put just about enough stories into his head," and packed them all off to bed.

· · ·

Colton and Preston left Evanston for Togwotee Pass after breakfast the next day, Colton hanging out of the window of Preston's pickup truck, "Whee-haw!" waving at his father and grandpa until Preston said, "Get your head back in here and wind up the window before we both die of cold." They drove most of the day and set up camp on about eight or nine inches of fresh snow in the late afternoon, Colton built a fire and Preston set up the tent. The boys ate quickly, cupping their hands around their bowls for warmth. Then they set their food and plates in a bag up over a bear pole and contemplated their sleeping quarters. All they had was a little nylon two-man and Preston was six foot four by then, even without his boots on. "Let me get in first and get settled," he said, disappearing into it with some difficulty.

"Holy cow, Preston," said Colton, crouching down and poking his head into his side of the tent, "there ain't no room for nothin' in here 'cept you."

"Lucky enough you're a runt."

"I ain't a runt." Colton squeezed in next to Preston, head to toe.

"Well, you ain't full grown yet, by any means," said Preston. "Least I hope not."

Colton put his hands behind his head and streched out as far as he could.

"Look kid," said Preston, "get your elbows out of my space, would ya?"

Colton kicked around in his sleeping bag.

"You sleeping with your boots on?"

"Dang right I am."

Preston sighed, "Watch your language, kid."

There was a long silence, Colton breathing hard into his hands and rubbing his legs together. "Holy cow," he said at last, "I'm freezing."

"That's life for you," said Preston. "Better get used to it."

• • •

Before dawn the boys were up and tracking through the freezing crust of snow but they didn't see much and it wasn't until nearly evening that they got close enough to a three-pointer they'd been following since early afternoon to do anything about it. Preston crouched low. "Shoot now, kid," he said. "Get a clean shot and shoot."

"Now?" said Colton. He was like an aspen leaf, shaking.

"Kid, you got to get calm and shoot. Shoot it now!"

Colton brought the gun to his shoulder, the buck raised his head, Colton breathed out the way Bill had taught him and squeezed the trigger, and the buck tipped forward onto his knees and then slumped sideways.

Preston stood up to his full height, nodded, and spat.

"Holy cow," said Colton. "Holy cow! I did it! I did it."

"Yep."

"I did it!" yelled Colton, dancing around in the snow.

"Now we got to get him dressed before dark or the grizzlies are gonna be having a teddy bears' picnic."

"Holy cow," said Colton, running behind Preston, but his hand was over his mouth and he was already puking before they got to the buck.

"C'mon kid, better give me your knife," said Preston. He pulled a piece of rope out of his pocket and tied the deer's front legs to a tree, then he propped open its back legs with a branch. He jabbed the point of the knife into the deer's breastbone and ran two fingers in front of the blade to the animal's pelvis, moving its entrails to the side as the knife slid down.

Colton leaned with one arm against the tree and hollered up more of the contents of his stomach. He wiped his mouth with the back of his hand. "Holy cow," he said. "I never could stand the sight of blood."

"I spilled enough of yours over the years. You think you'd be used to it."

"That's different. I don't mind so bad if it's my own."

By now Preston was sawing through the ribs.

"Man," said Colton, sinking to his knees and puking some more.

"You got anything left to puke up there?"

Colton waved one hand at Preston. "I'll be okay in a minute."

"We don't have a minute," said Preston. He cut out the deer's genitals and then he cut around the deer's anus and tied it off with a piece of string.

"Holy cow," said Colton.

"It's okay," said Preston. "It takes everyone a little differently."

"Holy cow," said Colton again, rubbing his eyes, "I guess so."

. . .

By the time they got back to the tent with the deer slung between them the boys were exhausted. They hung the deer up out of reach of bears and washed the blood off their shoulders, arms, hands, and faces. Preston built a fire and heated up some water for coffee and made Colton eat some crackers. Then they piled into the tent and zipped it shut on themselves.

"Oh man," said Preston.

"What?"

"This whole tent smells like your puke."

Colton hiccupped miserably.

"It's alright," said Preston. "Just so the grizzlies don't get wind of it."

"Holy cow," said Colton. "What bear's gonna want puke-covered cowboy for supper?"

And then there was a long silence until Preston said, "Colt?"

"Yep."

"That's a nice little three-pointer you got."

"Thanks," said Colton.

"Now next thing you need to do is break a mustang."

"Holy cow," said Colton.

"That'll teach you about everything you need to know," said Preston.

"Everything about what?" said Colton.

"Everything about everything," said Preston.

4

BILL'S PHILOSOPHY OF
HORSE BREAKING

Evanston, Wyoming

———————————

They fetched the wild mare home from the Bureau of Land
Management auction at Rock Springs for Colton's sixteenth birthday. She was desert-tough and sandstone colored, with a touch of
Arabian evident in the dish of her face. Maybe a few generations
back a rancher or, more likely, a ranch hand had gotten tired of an
Arabian's uptight tail-twisting ways and turned it loose on the
BLM to fend for itself. "Go tell it to the mustangs," says the
ranch hand, opening the gate, and the Arab, tossing its head in
serpentines of willfulness, making for all the high-dry land it
could pace between it and the boredom of confinement.

But a handful of generations later, any refinement of senti-
ment that may have made its way through the bloodline and into
her face had not stayed in this mare's spirit or her general physi-
cality. She had the sort of no-nonsense stride that indicated she
planned to make miles before dark. She had hooves of iron, the
shaggy remains of a winter coat on her like a moose, the startle
reflex of a jackrabbit. And because of her very first introduction to
the hands of mankind, she'd never be easy to catch.

She had been herd-torn in late spring by a helicopter. Rivers
of horses flat-out panicked over the sage and baked red earth
until fences funneled them into smaller and smaller corrals. The

shouting humans, the sting of fuel in the air, the prods and ropes hot in the air, the great clanging metal that surrounded her were a terror beyond all instinct—an unknowable enemy—the way the ground picked and moved off from under her feet, rumbling at terrific speed with all the horses crammed flank to flank trying to stay upright against the tug of gravity as the truck banked corners. And then, after the shapelessness of freedom, the deadening dust and distinction of captivity. There was no moving away from it, no matter how much she flattened her ears and backed up into the sky.

. . .

They paid a hundred and seventy-five dollars for her and she would have been cheaper if Colton and Kaylee had sat on their hands and let Bill do the bidding. "You're bidding against one another," Bill told Kaylee. "What are you hoping to pay for the thing? We ain't buying a racehorse here." But now the horse was his and Colton was so excited that Bill made him sit on his hands the whole way home. "So help me, you'll have us all off the road, you hopping about like that," he said.

"He-he-he," said Colton.

And the next morning before the sun burned the frost off the spring-sharp grass, Colton was out of bed, tugging on his Levi's and his cowboy boots. He pulled a T-shirt over his head and covered his hair with a ball cap and let himself out into the paddock, the old black cat at his heels. "Don't want to get kicked," he told her, nudging her with the toe of his boot back onto the porch. He roped the horse then and got her into the round pen with some difficulty, the mare plunging backwards and Colton tall and skinny and whipped about as a cornstalk in a high wind on the end of her rope.

An hour into it, Bill came out of the cabin, a plug of tobacco in his lip, a tin of Copenhagen where it had sun-bleached a pale circle in his jean pocket, the old black cat tripping around his feet.

"Move yourself," Bill told her, "you make a man's boots itch just to look at you." He pulled his cowboy hat down over his eyes and propped his boot on the bottom rail. Colton was coming at the mare with his arms outstretched, like he was trying to direct an airplane on a landing strip.

"I'm gonna touch this mare in half an hour," he told Bill.

Bill said, "Sure you will, son." He watched their feet, Colton's dusty boots stepping toward her, the mare's black hooves pacing away from him. The boy making advances, the mare's feet tense, ready to bolt.

"I got her under control," said Colton.

"Sure you do," said Bill.

Colton said, "Whoa now girl, steady," and took a step forward.

The mare almost flipped over the rail.

"Crap," said Colton.

The old black cat settled in a dusty nest, yawned, and closed her eyes.

The mare did a couple of laps of the round pen, checking the rails for a gap. She stopped at the gate, haunches bunched to the boy. Colton set a second lasso above his head then, eight beats in a circle like an eddy of silver water before he let it out and the rope went spilling over her neck. The mare felt the rope tightening and she startled back, her eyes walling white and her feet pounding, the rope pulling and pulling and running hot over Colton's hands until she burned right out of his grasp.

"Stupid horse!" said Colton, throwing his ball cap on the ground so that the mare startled like he'd just thrown down a coiled rattlesnake.

Bill said nothing.

"Okay." Colton picked his cap up and put it back on his head. "Have it your way."

The mare sighed and stretched her neck out, sniffing the ground where the cap had been. Colton took a step forward. The mare's head shot up and she snorted. "Don't make me rope you again," Colton warned, shaking his arms out loose by his sides. The mare took three bouncing steps away from him and stopped opposite the boy, her flanks pumping. When she stepped forward again, she caught her front hoof on the end of the lasso and her neck jerked up short. Her eyes went wide but she kept her nose level, the better to see, front, back, up, down—she was having a hard time figuring out where the danger was coming from.

"The boy'll get wore out doing that," Bill told the cat.

"You'll get wore out doing that," Colton told the horse. He lifted his hands up and came at her that way. The horse sawed back and forth against the rails, watching the boy out of the corner of her eye. At the last minute she blew up and belted past him before he could get a finger on her, spraying him with sand. "If I had a gun right now, I'd shoot you," said Colton. "Stupid horse."

That's when Bill said, "Try giving her a name. You gotta call her something more than 'stupid horse.'"

By now Kaylee was out at the round pen. She's pretty in a delicate Farrah Fawcett kind of way with flipped strawberry blonde hair and pale blue eyes a little worn at the edges by four kids and life in a high, dry climate. She put her arm over Bill's and he put his hand over hers. "How's it going?" she asked.

Bill said, "About right."

Kaylee smiled. She raised her voice and said, "Why don't you call her Cocoa?"

"What?"

"Cocoa," shouted Kaylee. "Let's call her Cocoa."

Colton said, "You can call her any way you like, but she's still a stupid horse."

"Watch your language, son," said Kaylee. She went back into the cabin.

"Dang horse," said Colton, spitting.

By now a pale layer of kicked-up dust reached waist deep in the round pen. Cocoa had soaked through the remains of her winter coat. She gave off a smell of ammonia and sweat. "If you'd just let me near you," said Colton, "this would all be over." He took a step forward.

Cocoa spat another lap around the rails, breathing hard.

"Stupid horse," said Colton.

So it went all the rest of that afternoon and Bill watched, saying nothing, until it was too dark to see. Bill figured maybe tomorrow or the day after, or maybe the day after that, Colton would get to touch her. And maybe a day after that, he'd get to throw a blanket on her back. And maybe a month or so after that he'd have her well enough under his skin to know that he was better with her than he was apart from her.

There are, Bill figured, some things best learned the hard way.

5

BILL AND COLTON

Evanston, Wyoming

———◆———

It would be a cliché and also not entirely accurate to say that Bill looks weather-beaten, because he doesn't look beaten by the weather, or by anything else. So it might be better to say that Bill is a man uncovered by weather—blown and rained and sunned and snowed—to the essence of himself, more and more perfectly grained with every passing year. Stripped of unnecessary flesh in this way, he hangs faultlessly on his own bones, so self-contained that he couldn't lose his fundamental nature even if everything else were lost. And maybe because of this, he bestows a perpetual half-smile onto the world, magnanimous under his black mustache, like he isn't really taking any of this as seriously as other folk do.

So, even thinking about it hard, it would be difficult to say that Bill looks anything but iconic because other terms ordinarily applied to someone's appearance—handsome or homely, for example—seem too trivial to apply to him. But it is easy to see how he inspired hero worship in Colton, and in everyone he has ever met, by the way he emanates soul. Tough-bound soul, like a monk. A high-altitude, big-sky, oil-drilling, saddle-bronc-riding monk who doesn't have any special thoughts on the matter of celibacy or God.

He's someone you'd never want to disappoint and it has nothing to do with anything he says, or even with anything he does. It's just the way he looks at a person from the window of his cherry-red pickup or from the back of his buckskin gelding with that custom-made half-smile on his face, like whatever you've done, or are thinking about doing, Bill Bryant's already seen worse, and anyway, the way he sees it, it ain't any kind of business of his what kind of soul you're in the process of choking down.

• • •

In Las Vegas once, Tabby, aged about ten, had to be physically pulled off a man who made a flip comment about Bill's cowboy hat in her hearing and, almost beyond belief, had actually reached up and touched it. For starters, if there's one thing you just *know*, it's to leave a man's cowboy hat alone. For another thing, Bill kept a couple of hundred dollars folded into the band inside of his hat against the event of an emergency. "Flat-tire money," he called it. Also, no one had ever spoken to Bill in anything but respectful terms as far as Tabby knew and she didn't see the benefit of it starting now, not on her watch.

"How much that hat hold, cowboy? Twenty gallons?" the man had asked, fingering the brim.

And before he could take another step, or say another ignorant word, or—God forbid—handle the hat any more intimately, Tabby was on his back, heels flailing, fists pounding. "Don't you touch my daddy's hat! Don't you talk to him that way!"

"Now girl," Kaylee told her oldest daughter, once the dust had settled and the man had been sent on his way, straightened out in the matter of how, or when, if ever, a gentleman talks about another man's cowboy hat, let alone touches it, "you got to learn. You can leave your daddy to fight any battles that need fighting."

"But he touched my daddy's hat!" cried Tabby, unrepentant, a fresh surge of indignation welling up inside her so that she had second thoughts about giving in so easily and letting the man get off lightly. "And I didn't even give him what he deserved. That man's still *walking*, isn't he?"

All this adoration from his family for a man who was home

only part-time the whole while the kids were growing up. "I'm not having my kids dragged around from patch to patch," Bill told Kaylee early on. So the moment she was pregnant with Preston he told her, "Pick a town you like. You and the kids are staying put till they're all grown. I'll chase the rigs and be back as often as I can."

In 1973, just before Preston was born, the Organization of Arab Petroleum Exporting Countries announced that they would suspend shipment of oil to countries that had supported Israel against Syria and Egypt during the Yom Kippur War. That suspension included shipments of oil to the United States. At the same time, a reserve of oil and natural gas was discovered just outside Evanston in southwestern Wyoming. Almost overnight the town grew from a quiet ranching and railway stop of four thousand to a city of twenty thousand. The oil work seemed steady and close, so Kaylee chose to move there. It's not usually smart to count on boomtowns to stay booming, but when the second oil crisis hit in 1979 (by which time Tabby had been born and Colton was on the way) Evanston solidified on the map: schools, a golf course, a recreation center, and a horse racetrack. And since then, through boom and bust, Evanston has always been less than a day's drive from any oil patch Bill has ever had to work in whatever combination of shifts the oil companies can dream up.

There have been some rigs Bill's worked on where he was gone for a month, back for a week. He's done two weeks on, two weeks off. He's done a week at a time. He's done flat out, day in, day out, until the hole was drilled. He's done pretty much every variation of time you can think of. And in time, first Preston and then Colton followed him onto the rigs. What hasn't changed is the company Bill's drilled for—for over thirty years he's drilled for the same company, but they still have him on their books as a part-time laborer, which makes it easier for them to fire him the moment he gets too old or too slow, or if he slips. And just

recently, some kid out of the head office saw a ten-gallon dis-
crepancy in a fuel tank filled by Bill and fired him on the spot. So
the next week, Bill was back in Casper submitting himself to
urine tests and physicals and safety talks so he could sign on with
a new company.

But none of this seems to bother Bill much. He looks at the
terms of his employment much the way most men think of
women or weather, as something beyond the power of his control.
Or like the way a hitch will shift from under your feet; day to
night, one week to a month, from a fortnight back to a week. It's
all down to someone with a computer in Houston or Casper or
Cheyenne typing you into a drilling roster on a desert or a high
plain they have never seen, and have no intention of seeing, as if
you're a megawatt.

What always stayed routine for the Bryant family was Bill's
time at home. If Bill had been on the night shift, he could be back
again in Evanston by early afternoon from Baggs, Big Piney,
Farson, all over Utah. And if he had been on the day shift, he
could be back at midnight or two in the morning and still up with
the birds so as not to waste any daylight on sleep. He left his boots
and his greasers on the porch by the front door and at some point
during the week he'd take them to a Laundromat with his
youngest daughter, Merinda. That was what they did for special
time—sit together in the Laundromat, not talking much, with
Merinda basking in the silence—while the work clothes fouled up
the industrial washer and left black, indelible smears on the glass
window.

Meantime there would be Colton at home, ready and waiting.
"I got the horses saddled, Dad."

"Let me get myself a cuppa coffee, son, and I'll be right out,"
said Bill, though it had been twenty-four hours since he had seen
the inside of his eyelids for any length of time.

"I got your coffee for you right here," said Colton, handing Bill
an orange mug, thirty-two ounces from the Maverick Gas Station.

So all year, every year after he learned to ride and saddle a horse, as long as the snow was still shallow enough that a horse could clear its belly over it, Colton and Bill headed up into the hills on Bill's time off the oil patch, sometimes for a few days at a time. That August, after he'd had her a couple of months, Colton persuaded Bill to take a video camera along to show what a good trail horse Cocoa had become, but Colton wasn't counting on Bill's gelding blowing up at a spooky herd of elk, leg-crashing through the aspen tree deadfall, eerie as ghost-creatures the way they vanished.

Not that the gelding could unhorse Bill, who, after all, has a saddle-bronc rider's sense of balance, but the video doesn't show too much of Cocoa. You can see her and Colton disappear into a grove of trees, Colton all camera-silly and blowing his nose loudly enough to blast his brains out, and then there's an elk's rump and another elk in the shadows, and after that the video shows what it's like to be on a bolting, bucking horse from the rider's point of view. The soundtrack is pretty telling too. No one says much of anything but you can hear Bill say, in a calm drawl, "Dang horse."

And you can hear Colton laughing, "He-he-he," like he just caught the punch line to a joke, which, in this country, a wreck off a horse almost always is.

6

In the Beginning

Wyoming and the West

———— ❖ ————

So Colton was born with horses *and* oil in his blood like his father before him and his grandfather before that and maybe his grandfather's father before that. Who knows, because Wyoming is repeopled every time there is another oil boom, transience refreshed and history forgotten. People arrive in Wyoming on their last tank of gas, no way to chicken out at the last moment and go back to whatever it is they were running away from, weighed down with a new heartload of all the old reasons for starting fresh. And then the boom's over and the brokenhearted leave, and it's all unpeopled trailer parks and motels with their peeling backs to the long set of the afternoon sun. The wind blows the same anyhow, boom or bust, although more hollow with less people there to hear it.

One time, Bill Bryant's mother helped him count up all the places they had lived while he was growing up and they figured thirty-two places in fifteen years all over New Mexico, Wyoming, Utah, North and South Dakota, Colorado, chasing down oil pays, one trailer park or motel room at a time, four shabby walls between them and all the bad-tempered weather the West has to offer, and then someone would hang a sign on the edge of town, "Last one to leave, turn out the lights," and they'd be off to some-where new until it all started to look like one place and a new year felt a lot like the old one.

Even in the good times, it was a struggle because the price of everything went through the roof during a boom. A hundred bucks a month for a room, instead of twenty, a bag of groceries worth more than a bag of cash, that kind of thing (ten times that, these days). And people didn't and don't seem much happier than they had been before. Campers and trailers settle with an air of temporariness on the sagebrush desert; bars without windows so you can drink all day and night without being reprimanded by sunlight. A fair number of people show up with Oklahoma, Ohio, or Louisiana plates, fail the first piss test they take, and swell the numbers at the food bank worse than in a bust. There's a lot of what keeps you up and enough of what puts you down: methamphetamine, drive-thru liquor stores, a few strip clubs and plenty of porn. Even the names of the places hosting the booms have a jokey, ghost-town-like quality to them: Rock Springs, Wamsutter, La Barge, Casper.

. . .

By the late sixties, when he was just fifteen, Bill left his parents and was on his own, breaking colts for a rancher and riding saddle broncs at weekend rodeos for a little extra cash, walking away with more than the occasional belt buckle. Barely man enough to grow a stain of mustache on his upper lip, he was ropey, not yet fully grown into his body, but what he lacked in bulk and strength he made up for with a tolerance for pain and an uncanny, you might say physical, understanding of time. He knew in his body—the way blood started to pool in his neck, the way air was already leaving his lungs, the way his weight was being lifted out of his ankles—the exact shape of eight seconds. Eight seconds is all a bronc rider has from the moment the chute blows open until the bell, until the pickup man comes galloping up behind you and you can lean across the air, wrap your arms around the pickup man's waist, and allow the bronc to buck free away from your legs.

But when he was twenty-two, the ranch he was working on sold and Bill had to look for other work. Then as now, the best-paying jobs in the West for someone without a college degree were out on the oil patch. So Bill packed up his saddle and belt buckles and folded up his mattress and sleeping bag and followed his father onto the rigs. But he kept with him the impeccable timing he'd learned riding saddle broncs, and he never lost his horse sense, moving deliberately around that massive, sometimes unpredictable equipment like he didn't want to startle something or get kicked. What he passed on to his son was a desire to be just like him. What he could never teach Colton was a saddle-bronc rider's trick of slowing down time until you knew the shape of it, until you could possess it, until it was yours to stretch out or shrink—knowing that eight seconds is both short enough to hold on to and long enough to get yourself killed. That's the full, fat poetry of eight seconds.

7

CATTLE DRIVE

Near Evanston

———◆———

At dawn, in mid-September—the fall equinox—it's seventeen degrees in the valley a few miles west of Evanston. If there had been enough moisture for a frost, there would surely be a crisping of white over everything now. As it is, the cold is just dry as dust, settling on cattle and sagebrush, earth and skin, cracking everything. Colton and Bill are heading into the heart of the valley with a pickup and horse trailer. Behind a wind-battered banner that advertises a fall bull sale, testicle festival, it says, there's a ranch that follows a broken, eroded creek. On either side of the creek there is a flush relief of cottonwoods and a green vein of vegetation follows the wetlands up and up, eventually swelling into the Salt River Range. It's an old ranch for around here—a hundred years of permanent settlement or more—and as such, it's the kind of ranch that has stories creaking out of the walls of the old barn that start with the words, "It was the year the snow came up over the roofs . . ."

They unload Cocoa and Bill's hotfooted gelding, already puffing himself up with imaginary enemies for all the world like a Texan politician. The sun is just making long-fingered shadows through the sage on the hills. Colton's whistling to Cocoa while he checks her hooves, puts her bridle on.

"What you got there?" asks the rancher, thumbing his cowboy hat toward Cocoa. The rancher is a very tall man, and wind-whipped thin, compensating for his height with an apologetic hunch of his shoulders.

"You like her?" asks Colton. "I got her for my birthday in June."

"It's got four legs, anyhow," the rancher agrees. He steps into his stirrup and lowers himself into the saddle, rolls a little paper cigarette that he sticks onto his lower lip.

"She's a mustang off the Red Desert," says Colton.

The rancher lifts his left eyebrow a fraction. "That a fact?" He lights his cigarette and squints at Cocoa through the smoke. "Well, I'll be . . ." he says.

Bill is already in the saddle. He turns his gelding to face in on Cocoa, cowboy style, to catch anything that blows up. Colton vaults up onto the mare. "You'll see," he tells the rancher.

"No doubt," says the rancher, gripping the end of his cigarette between his teeth. He pulls the brim of his hat down another fraction of an inch and then he lopes ahead. "Watch for badger holes," he shouts over his shoulder.

By midafternoon it's over seventy degrees and the sun is unfiltered. The horses are sweating and there's a sting of salty heat coming off their coats. The rancher is working his mount right up into the herd and shouting, "Hup! Hup!" and the cows are bawling and the dust tastes of manure and sweat. The tags in the cows' ears shake blue, yellow, orange, like wind chimes. Most of the six hundred head are moving complainingly toward the home paddocks along a red dirt road but a few are trying to thread their way back to their old grazing grounds with the dogs worrying at their feet.

From the sky, this little cattle drive would look like an ants' parade with a child's careless stick drawn through the middle. The rancher, Bill, and Colton are all small and beetling and nothing, hardly moving at all, and the cattle are in rivers of stupid confusion milling mostly forward. From the sky, against the immense, witnessing bulk of the Salt River Range, all this seems pointless. All this work, this noon-flattened light of unreasonable cows and dust and sweat seems as if the riders have come nowhere at all and as if they can never go anywhere.

When they come to the end of the spring, Bill nods at Colton. "Go back and make sure we didn't leave anyone up there."

So Colton turns Cocoa back and trots up the stream one more time and Bill keeps going forward with his puffed-out gelding. Then a noise erupts behind him. Bill looks back and sees Cocoa coming at him at a flat-out neck-stretched gallop, the reins flapping loose on her neck. Colton's leaning all the way back in the saddle, arms outstretched, face thrown toward the sky, the back of his head connecting with Cocoa's rear. The sun is behind them and the air is shocked golden. Colton like a crucified Jesus on a horse. "The Injuns are coming! The Injuns are coming!" he shouts. "I've been shot in the heart."

Bill's gelding frights itself half to death in a tangle of willows and Bill says, "Dang horse . . ." and spits. He needs to wipe his eyes. "You could get yourself hurt like that," he tells Colton.

Colton picks up the reins again and puts his ball cap on straight. "Maybe they'll put me in the movies."

"As a dang fool?"

"As one of them cowboys that gets shot in the heart."

"With a nose like that?" says Bill. "They'll need a pretty wide screen. Make that a wide-as-all-heck wide screen."

"He-he-he," says Colton.

"He-he-he," says Bill. He pulls his cowboy hat down over his eyes and spits. "C'mon, son, we got some cows to move."

8

GOOSE HUNTING WITH
JAKE, COLTON, AND CODY

Near Evanston

When they weren't messing around with horses or guns or video games or whatever else snagged their flighty attention, Colton, Jake, and Cody had spent a great deal of that summer and a good part of the fall messing about with Cody's little Mazda pickup. They'd stripped it and straight-piped the thing and put a glass pack on the tail but that little Mazda still didn't sound any louder than a pissed-off lawn mower. Nonetheless, there they were, trying to muscle up an impression of invincibility on the dirty-iced streets of Evanston, the three of them packed in the front seat, shotguns and a dinky beat-up spinning rod in back, cans of Copenhagen on the dash. Under their legs, they'd stashed skinning knives and Jake's swamp-rotten waders, the latter of which were now giving up scents of old frog and last season's mud. Neil Diamond was on the player like it was summer forever and no one was ever going to grow old and luck and love were on the side of all God-fearing boys in blue jeans.

Almost none of this was true:

It was early November and the yellow light was sickly with whatever was coming to them. Town, stripped of summer ice-cone stalls and fall leaves and not yet covered with snow or Christmas lights, was depressing the way a hangover is depressing, a low-grade headache of a place that must have seemed like a good idea at the time. The old railway buildings and Chinese laundries just north of downtown seemed bewildered, unwanted from another age. The low new buildings in town—EZ-tanning and payday loan-type joints so tenuous that they hung their names out on plastic banners instead of anything permanent— looked cheap and too thin for the weather. But none of this mattered to the boys. They had guns and Mountain Dew and southwestern Wyoming seemed to be there for whoever was alive enough to take it.

. . .

The reservoir isn't much to speak of, unless you love that sort of thing. Just a big dirty hole of water on the outskirts of town, but the boys knew it like it was heaven with a side of freedom. "Money talks, but it don't sing and dance and it don't walk," the boys sang.

"Honey's sweet, but it ain't nothing next to baby's treat."

"Money talks, but it don't . . ."

They bounced into the parking lot at the reservoir. There were no other cars around. The outhouses had been closed for the season, garbage cans turned upside down against the anticipation of inevitable snow, the picnic sites scraped up. A sign asking visitors not to litter had been buckshot in the belly. The boys piled out of the Mazda, Cody and Colton ahead of Jake, all of them crouched low, soldierly, shotguns across their chests as they jogged toward the reservoir on a thin layer of crunchy snow. The sun had given up trying to break through the dull press of sky and had slid into someone else's tomorrow. The cold had settled on the land like an endurance test. I don't suppose the boys really believed the geese would be roosting on the ice-covered reservoir, or that there would be so many of them this late in the year, but there they were—dozens of geese balanced as still as decoys on a half inch of ice. "Holy crap," said Cody.

"Shhh," said Colton. He spat and a yellow stream flew from his mouth. A bit of tobacco fell on his lip.

"What do we do?" asked Jake.

"Open fire, boys," said Colton. He nodded at Jake and Cody. "Now."

And then, all at once, the three boys had their guns to shoulders, unloading on those geese as if they thought it likely the birds would try to fire back. The geese startled up, struggling to get airborne on the thin, frigid air, skidding on the ice, everything heavy with a terrible kind of possibility and slowed up like time had snagged itself on the cold. The boys pumped shot after shot down onto the reservoir and when the smoke cleared there was just one goose left on the ice, every other bird airborne, honking distress into the pale grey evening. Jake lifted his gun and fired one more shot. That lonely icebound goose gave a startled jerk, but kept waddling toward the middle of the lake. A tiny black hole of open water gaped ready for it.

"Did I hit it?" asked Jake.

"Shoot again," yelled Colton.

But at that moment the goose skidded one final step and fell over dead just within reach of the hole and there she floated.

Then Colton was on his toes dancing like someone suspended by ropes from the sky, all arms and legs. "Your first goose, man! It's your first goose!" His knees pumped up and down. "It's your first goose!" And then he stopped dancing suddenly and reached down his pant leg. "Sonofa, I forgot . . ." He fetched up a can of Mountain Dew.

"What were you doing with that in your rods?" asked Cody.

"So it won't freeze." Colton pulled the tab and took a sip. "Cheers, boys."

"Holy cow," said Jake, looking out at the dead creature floating in the little hole of water in the middle of the iced-up lake. "I wish I had a dog right now."

"Here," said Colton, handing the soda to Jake. "Hold this."

"What you doing?"

"The redneck retriever."

It was almost completely dark and Colton had already tried pushing logs across the reservoir after the goose, throwing rocks at it, and, finally, slogging into the water in Jake's leaky extra-large waders, breaking ice with his chest as he went, the dinky little spin rod from the backseat of Cody's pickup flailing ahead of him, trying to catch the goose on an oversized lure. From the Mazda, where they had their feet up against the heater, Cody and Jake shouted directions out the window, but whatever they said, the fishing line wasn't very long and the lure kept landing close enough to Colton that the boys could hear the plunk as it hit the ice. Then as Colton got out to waist-deep water, "You can't swim!" Jake reminded him.

"It's not so deep," Colton shouted back. But a few more steps and the water started to come over the top of the waders.

"You can't swim!" Jake repeated.

"He can't swim?" said Cody.

"Not so as you'd notice," said Jake.

"Holy crap," said Cody.

"Dog-paddle," said Jake.

"This is how you see on television about people dying," said Cody.

"He'll certainly freeze," observed Jake.

"Most definitely," said Cody.

The two boys watched their friend for another few minutes. Then Jake said, "You hear about the kid they found up in Sublette?"

"Nope," said Cody.

"Cops found him in the desert with his head in a badger hole."

"Doing what?" asked Cody.

"Dead."

"Holy crap," said Cody.

"They say he'd been there for months. Sunbaked from the ass down."

"With his head in a hole?" asked Cody.

"Yep."

"Holy crap."

"Yep."

"Crazy freakin' sonofabitch," said Cody, looking out the window.

"Should we go fetch him?" said Jake after a few more minutes.

"Colton?" said Cody. "Fetch Colton? Since when do you think he'd listen to us?"

Colton tried a few more casts from chest-deep water, but he wasn't even close to catching the goose. "Okay! I'm coming back," he shouted at last.

By the time Colton climbed back into the pickup he was most definitely starting to lose higher functioning—his systems shutting down from hypothermia—and the way Jake tells it, Colton didn't have an excess of higher functioning to lose in the first place.

"Man, you're frozen like a frozen thing," said Cody. "You coulda died out there."

Colton was too bunched with cold to speak, his lips pressed together and blue.

"We got to get these waders off you," said Jake. "You're soaking, man." So Colton was wrestled out of the waders. Then Jake sat rubbing Colton's hands and Cody turned the heat up and complained about the pickup getting so steamy it was like the freakin' Amazonian jungle—all they needed was them snakes and bugs and freakin' Injuns with them bones through the noses—and this went on until Colton started singing between chattering teeth, "If I should die before I wake, feed Jake . . ."

"Oh brother," said Jake. He looked at Cody. "His brain must be turning back on. He's being a retard."

"Think," Colton said, wringing his socks out with shaking, white hands. "How we gonna get that goose?"

"Can we forget the goose?" said Jake. "You're about froze to a standstill."

"I ain't gonna forget your goose, man. It's your first goose. And anyway, my dad's gonna tan my hide if he hears I left a dead goose out there."

So the boys sat in the truck listening to Cozy Country 106.1 FM for another twenty minutes. Then Cody said, "Happy yet, Colton? It's completely freakin' dark. Can we go home now?"

"Can we come back tomorrow with a bigger rod?" asked Colton. "Maybe if I just had a bigger rod."

"Whatever you need to do."

"It's Jake's first goose, man."

"I know," said Jake. "But there's no sense drowning for a dead goose."

Colton gave Jake a look like he was thinking maybe there was.

9

JAKE

Utah

———◆———

One afternoon when Jake was twelve years old, living with his parents and five siblings on the family ranch in Utah, he pumped three shots into his father's shotgun and held the barrel up to his elder brother's head—the brother's teasing had been getting on Jake's nerves—and he said, "You touch me again and I'm going to blow you away and chop you up and put you in a salad and eat you and no one's gonna know where you've gone." This was after he'd already called the teacher in first grade a kootchie-snatcher, flipped the bird at the lunch lady every schoolday for six years, threatened half the kids in his class with acts of such imaginative violence that his parents said to the teachers, "We don't know where he's getting this stuff."

Eventually a shrink in Salt Lake City got it out of the boy. There had been something bad going on for some time. It had to do with some people Jake and his siblings called "Aunt" and "Uncle." And whatever it was they'd done, it was bad enough to make an attempting murderer out of a twelve-year-old boy. When Jake's father found out the extent of it, he held a pistol up "Uncle's" rear and told him, "You come near my family again as long as any of us shall live and I swear I'll unload this right into the middle of your world. You get me?"

But you can't shoot the ass out of everything bad that happens to your family, and by the time four of their six kids had gone to a psychiatrist for a couple of years to recover from what it was "Uncle" and "Aunt" had done to them, Jake's family had lost their ranch. This was in the early nineties, when beef prices went from something like a buck ten to eighty-three cents a pound overnight. And on top of that, Jake's dad had been all but paralyzed in a combination of accidents; he'd broken his back falling off a semi truck when he was loading feed bags and then rebroken it in a more recent horse accident. Plus, he'd bust his heart over everything that had happened to his kids and there wasn't a decent heart doctor for five hundred miles. It was scenic here, true, with the canyons nearby, red and impressive, but scenery doesn't feed the cows and even if it did, cows don't hardly pay for the groceries let alone the specialists' bills.

Jake eats under pressure and the pressure of everything that had happened so far just about killed him at an early age. He was fourteen years old and he weighed nearly three hundred pounds. So it was not surprising that one morning Jake's dad looked over the breakfast table at his family and said he hated to see it come to this, but the ranch had to go. Within a month, he sold the place to a man who could afford scenery and didn't much mind about beef prices. And with the money from the ranch, Jake's father bought a Subway franchise in Evanston.

Jake lost fifty pounds in the first six months after leaving the ranch. This was all before Jared Fogle, the "Subway Guy," lost two hundred and forty-five pounds in one year eating a small turkey and a large veggie sub every day with a diet soda and a bag of baked potato chips. In any case, it wasn't a Subway diet that lost those pounds. It was the fact that Jake didn't have to wake up and go to bed right in the rotten heart of the memory of what had been done to him. He still weighed two hundred and fifty pounds, though, and that's a fat kid, no matter how you look at it. And how to explain to his classmates that childhood trauma made him eat? You try getting sympathy in Wyoming for trauma—

childhood or otherwise. Cowboy up, cupcake, everyone has their freakin' problems.

So outside of the shrink's office in Salt Lake City, Jake kept his mouth shut about what had happened to him and he tried not to think about it too much. And pretty soon he took up deer hunting.

10

JAKE

Evanston, Wyoming

And then, a few years after the family moved to Evanston, "Uncle" showed up in town, with who knows what on his mind, driving back and forth past the Subway shop where Jake's sister was working the counter. "Uncle" phoned the shop and when she answered, he said, "You can't hide from me. I can see where you are."

Jake's sister screamed.

"I'm right in front of your eyes," said "Uncle."

Jake's sister looked out of the window, saw "Uncle" driving slowly by the spring-crusty banks of snow. Then she looked around for the quickest exit and came to the panicked conclusion it might be in her own hands. By the time Jake and Jake's father got to the scene—locked and loaded—"Uncle" was gone but Jake's sister had red stripes on her neck from where she'd tried to strangle herself with the phone cord.

Jake put his hands on her shoulders and made her look at him. "You listen to me," he said. "I'm gonna take care of this for you. You won't ever have to worry about him again after today. Okay?"

Jake's sister nodded.

"I mean it," said Jake.

Jake's father said, "I warned him."

He and Jake piled back into the pickup and drove around the streets of Evanston for some time with rifles across their laps, ducking down alleys, stalling up on dead ends where the plow had left icy, grit-crunching banks of snow, circling up by the state mental hospital where sure as hell the sonofabitch belonged. Then they took their search out of town, but what existed outside city limits was so much of everything wide open, a person could evaporate out here, hide in the snow-blown creases that make up the endless quality of Wyoming's open spaces. You might look all your life and never find a man, which is what makes it such perfect outlaw country. So eventually they turned around and came back into town.

"I guess he won't be coming back," said Jake's father.

Jake didn't say anything.

"I think we showed him," said Jake's father.

"He ain't a coyote," said Jake. "He ain't that smart."

. . .

By the time Colton got to Jake's house, Jake had made up his mind.

"You missed school," Colton said.

"I know," said Jake, sliding a fistful of ammo into a duffle. He had two guns on his bed.

Colton said, "What you doing?"

"Nothin'."

"Going hunting?"

"You might say that," said Jake.

"Without me?"

"Yep."

Colton thought about it for a moment. "What's in season?"

"A couple of assholes," said Jake.

"*People?*"

"Yep."

Colton jumped to his feet. "What are you doing?"

"I said already, I ain't telling you."

Colton watched Jake stuff a camouflage jacket and a black T-shirt into the duffle. "You got to tell me," he said.

"It's nothing you want to know."

"It is."

"No it ain't."

"Okay. Then I'm coming with you."

"No you're not."

"Jake, if you're gonna kill a man, at least tell me what he did to you."

Jake sighed. Then he said, "Alright, but you're not gonna like it. You're not gonna want to know me after this."

"Sure I will."

So Jake told Colton and Colton sat on the end of the bed getting whiter and whiter and with his hands over his head as though he was afraid it would blow off, and when Jake had finished he said, "Holy crap, Jake."

"See, I told you."

"See what?"

"You wouldn't want to know me."

"Not true," said Colton. "It's just," he said, "I mean, I guess I've had a pretty sheltered life. You know. Nothing like that ever happened to me." Colton tried to think of something that had happened to him and came up with nothing much. "Worst thing that ever happened was when I nearly froze to death getting your goose . . . Oh, and another time when I was a kid I put my fist through the living room window on Merinda's birthday 'cos I didn't like that she was getting all the presents and I wasn't getting nothing. That most definitely hurt. And . . ."

"Colt?"

"Yeah?"

"You finished?"

"Sorry."

"Okay."

Colton bit his nails and frowned. "It's just," he said. "I just don't know . . . I don't know what to do."

"I do," said Jake. "I know where they live. I'm gonna drive out there and I'm gonna put an end to this."

"Oh," said Colton.

"What do you suggest?" said Jake.

Colton said, "Most definitely that's a good question."

"Exactly."

"But I don't think blowing them away is gonna fix nothin'."

"Oh yeah?"

"Yeah." Colton looked down at his hands. "I mean, I'd shoot the sonsofbitches for you myself if I thought it would help. But it ain't gonna help, Jake. What they hurt, I don't think you can fix it with guns."

"It's worth a try."

Colton shook his head. "No, it ain't. It ain't worth going to prison for 'em, Jake."

Jake gave Colton a look like he thought maybe it was.

"No," said Colton.

Jake sat down on the end of his bed. "So what? I sit here doin' nothin' like a great big pussy?"

"Yep," said Colton.

"Since when were you all about forgive and forget?"

"I dunno," said Colton. "Since forever, I guess."

"And what if he comes back into town again like he just did?"

"Then," said Colton, "we'll deal with that when the time comes."

Jake looked at Colton long and steady and then he buried his face in his hands and cried until tears poured through his fingers. Colton sat where he was watching for about as many minutes as he could stand it then he said the only thing that would come to him: "Don't make me have to do my happy dance, Jake. You're gonna make me have to do my happy dance." And then he jumped up and started jerking around the room like a puppet controlled from the sky, knees lifting halfway to his ears, turning his back on his friend so he could wipe his eyes.

"Oh crap," said Jake. "Holy crap, Colt."

"He-he-he," said Colton.

. . .

After that, nothing much was ever said about the matter, though Colton knew, and once in a while if he caught a certain look on Jake's face he'd say, "You gonna make me have to do my happy dance," and it made no difference if they were in Ace Hardware or Porter's Fireworks or the middle of the rodeo or a restaurant, he'd up and dance, his knees coming up to his waist, his hands all around his ears.

And Jake would say, "I'm happy now, Colt. Would you quit it? I'm very happy now, you crazy freakin' sonofabitch."

II

JAKE AND COLTON

Evanston, Wyoming

—————◆—————

The fall after "Uncle" came to town, the fall that Colton was seventeen, the educational authorities in Uinta County put Jake in a special school for dropouts in a last-ditch effort to get him through high school. Jake's problem was that he had taken so much of his early education getting into fights with the other kids and flipping the bird at the lunch lady and abusing his teachers and generally being disruptive that he hadn't absorbed as much as he needed in order to graduate. Colton had dropped out already, just plain quit going to school one day. Colton's problem was that he couldn't take the Kmart cowboys and their incessant teasing for one more day. "Retard!" they still called him and all because he couldn't seem to keep his brain still long enough to keep caring about a sentence from its beginning to its end. "I lose interest," he told his mother. "I sit down and the book is all open and everything and then I look at the words and I can read them alright, but what's the point? They ain't *doing* anything. So I get to thinking about camping and hunting and Cocoa and the other one hundred and one thousand other things I'd rather be doing than sitting down staring at a bunch of words and I just about hop clear outta my skin it makes me so crazy. I want to be up and doing and *outside*."

"You need your high-school diploma," said Kaylee. "You won't get work anywhere without a high-school diploma. Not even the rigs'll take you now days without it."

Tears ran down Colton's face, "I know," he said.

"What is it, Colton?"

"Nothin'."

"Nothin' doesn't make a grown boy cry."

Colton wiped his nose on his sleeve, "It ain't nothin', Mom."

"C'mon, son. You gotta tell me."

Colton swallowed and rubbed his knuckles in his eyes, "It's the teasin', Mom. They ride me all day and never let up. Never even once. I pretend like I don't care, but I care. I care so much I can't sleep at night."

Kaylee made her hands into fists, "You just have to ignore what those kids say, Colton. You can't let them ruin the whole rest of your life."

. . .

Then Colton found out that for twenty-five dollars he could enroll at the special school where Jake had been sent, so he borrowed the money from Kaylee. "It's a surprise," he promised her, pocketing the check.

Kaylee wiped her hands on her jeans. "Don't you be buying me more flowers, son, you had to pawn your boots last time you did that."

"It ain't flowers, Mom. It's better'n flowers."

An hour later, Colton came back with a certificate to say he would be starting school the next week. "I'll get you that high-school diploma," he told Kaylee. "Maybe I won't be a rocket scientist, but I won't be a nobody neither. You'll be proud of me, Mom, just wait and see."

So they were like a little club of dropouts. That's how Jake explains it. With a few other boys whose names sound as if they complete the cast of characters in a modern Western—Cody, JR, Chase, Jake, Colton—they hunted and shot geese and Jake taught Colton to fish and they went camping and roasted just about a ton of marshmallows. They drove a lot too, filling up their tanks with the money they pooled from various part-time jobs at fast-food joints and car washes and the pawn shop. The boys didn't think anything of driving three hundred miles or more just to see what was there and to buy a burger somewhere new. This was the nineteen nineties when the conquered West was barely a hundred years old and when it was still full of a kind of gunshot, hard-won innocence and broken promises and open roads. Sure, there was hardship that went along with all that, and drought and violence and what have you—there always was and there always will be—but there was also a sense of freedom back then, like the freedom of adolescence right before you have to grow up and get really damaged in ways that never, ever heal.

"Look, man," Jake said, "we get through this stinkin' book and then we can go fishin'."

"Why can't we go fishin' now and read books later? I bet they're jumping like crazy right now."

"Because," said Jake," I promised you I was gonna get you through high school and I ain't a promise-breaker."

"So what? What's the worst thing can happen?"

"Here man, have another Mountain Dew."

So Colton cracked another soda and creased his forehead and said, "Sonofa . . ." and his mouth went tight with concentration. And Jake sat next to him playing video games and looking over once in a while at Colton's books and saying, "No, Colt, it's got to *equal*. That's why it's called an equation. So help me, are you trying to be stupid?"

"I ain't nearly as stupid as this stupid book."

Jake wrote some figures on the page. "Does that help?"

"Nope."

"Yes it does, look," and Jake wrote another few numbers and said, "See? It all adds up."

"Cocoa's going to be wild as a freakin' cat if I don't take her out for a ride, you know."

"No she won't."

Colton sighed and went back to his studies for another half an hour and Jake kept on playing video games and correcting Colton's mistakes and sending him back to do more work.

"I know," said Colton suddenly, slamming the book shut and standing up. "I got a better idea."

"What's that?"

"Let's go shoot bunnies."

"No way, Colt. You finish this first."

"Sonofa . . ." said Colton.

"Eight more questions," said Jake, "then we can go shoot anything you like."

"Sonofa . . ." said Colton again. He picked up a pencil. "When I'm dead everyone'll be sorry they made me waste so many hours in school."

"No they won't," said Jake.

"You watch," said Colton. "They most certainly will. Someone will mention it at my funeral. They'll say, 'We never should have made the poor boy do so much school. What a freakin' waste of his time it was.'"

12

RUNNING FREE

Near Evanston

All school year, Colton spent hours staring out of the school's windows at the mountains. From October until February, he watched the snow pile up, the dark close in sooner and sooner, and then he watched the bright spring sun come and the days lengthen and winter recede, and he started planning where he was going to take everyone camping as soon as the snow crept back off the peaks. By mid-March he was already telling everyone that what he wanted for his eighteenth birthday was a proper canvas hunting tent, six-man, with room to stand up in and swing an elk if he liked.

School got out a few days before Colton's birthday in June and he was a wreck of anticipation. On the morning of the tenth, he was up before it was light and sitting in the dining room when his family got up for breakfast, and within five minutes of getting the tent—better and bigger than he had imagined, with canvas pockets to serve as little storage areas and tall enough for even Colton to stand up in—he had arranged a camping trip, and by late afternoon Bill, Jake, and Colton were in the mountains with the tent and the horses and the marshmallows, just exactly as Colton had been imagining for the last six months.

Before dark, Jake and Bill fetched buckets of water from the lake for the horses and Colton stayed in camp to put the pegs in the new tent, and, since he could always be counted on to burn down half the neighborhood given half a chance, Bill put him in charge of making the fire. "See if you can do it with a little less tinder than you did last time, son," said Bill.

"Yeah, you pyro," said Jake.

A couple of years back, on a hunting trip, Colton had got back to camp before the others, having not seen anything but the flashing tail end of an elk since before the first river crossing. He waited in camp alone for a few hours—until after dark—and then he got to thinking and he decided maybe the others had got lost and it might be useful if he built a fire to guide everyone back, and by the time Bill and Preston found him, Colton was running around a wildfire with green branches torn off a baby pine and a good half an acre burned. "Holy cow," said Colton when he saw his brother and father.

"Keeping warm enough, son?" said Bill.

. . .

Jake and Bill came back from watering the horses and Bill put stones in a circle around the edge of the fire to boil water. "Nice to see you left some wood for the rest of the world," said Bill to Colton. The boys pierced sausages and then marshmallows on the ends of sharpened sticks and Bill made cups of strong black coffee. They ate their meal on tin plates on their knees, propped up against saddles, saddle blankets shrugged over shoulders, hats pulled down over eyes, cups of coffee by their feet. The heat from the fire released last summer's scents—sun and sweat, green grass and manure, elk and old blood. Coyotes were shouting victory and love across the valleys to one another. An owl asked, "Who? Who-who?" The boys didn't say very much, mostly because Bill didn't talk much, and the boys wanted to be as much like Bill as possible.

But at last Colton wanted to find a way to say how happy he was so he said, "Too bad you can't do this for a living."

"Wouldn't earn much," said Jake.

Colton leaned forward and poked his stick into the fire so that a shower of little sparks lit up around his face. "I reckon I'll be out on the rigs soon enough." He speared a marshmallow onto his stick. "Meantime I'll be out here roasting marshmallows." He looked up and grinned. "How's that for a plan?"

"You'll get eaten by a stray grizzly," said Jake.

"If I should die before I wake," sang Colton, "feed Jake . . ."

"Oh brother," said Jake.

Colton gathered the tin plates then and washed them in cold lake water and put everything that had been used for cooking in a bag and slung it up between two trees out of reach of bears. Bill stood up and stretched and tipped his hat back to look at the stars one more time. "You'll want to hobble her tonight before you turn in," he said to Colton.

"She ties regular," said Colton.

"She's not a regular horse."

Colton smiled. "Well I guess I know that."

Bill spat and pulled his cowboy hat down over his eyes.

Colton said, "You'll see. She'll be fine."

Bill sharpened his shoulders to the sky and disappeared into the tent.

. . .

Colton woke up next morning, rubbed his hands over his face and pulled on his ball cap. Jake was still asleep with his sleeping bag pulled up to his eyes. Bill's sleeping bag had been rolled up and left at the end of the sleeping mat. Colton looked around for his boots and tugged them on over his socks and then poked his head out of the tent. That morning was brightly frostbitten and clear all the way from here to as far as the eye could see. There was a lacy netting of mist coming off the lake, all secretive with what it knew about water and air and the difference between the two. The redwing blackbirds were shrilling, frogs had started chorusing, squirrels were chastising each other. A raven was making a noise as if two pebbles were dropping around in its throat. Colton threw a stone at it. "Keep it down, would ya?" he said. The raven gargled. Colton cocked his fore and middle finger at it. "Pah!" he said. "You're dead."

. . .

Just the other day, Colton had walked into the kitchen all strapped about with ammo and rifles and handguns, grinning like a dog on a hot day. "You look like Rambo, you know that?" his mother told him.

And Colton, hands loose by his sides, cupping the air around two holstered pistols on his hips had said, "You think?" pleased with the idea.

"Colton," said Kaylee, "son, that's not necessarily a good thing."

But judging from the portraits Kaylee made him sit for, Colton was built more along the lines of a very young Sean Penn, if you're talking any kind of movie star, right down to Sean Penn's impression of deep hurt lodged early somewhere far behind the eyes. Although it's hard to be completely sure because Colton didn't photograph well—maybe it was a combination of the scowl he liked to give cameras and the nose, too long to begin with, broken four times by Preston in sibling dust-ups settled with fists. "It's great to be able to wake up and smell the coffee . . . in Brazil," he used to say. And he had a goatee, the kind that has no author-ity on a face, and it gave Colton, even as a man, an impression of eternal adolescence. But in life he was six foot two in his bare feet, six four in cowboy boots, built whippy as baling twine but with wide shoulders and with a walk like he had never really found the difference between sky and earth, paddling all of it underfoot. And he took your breath away with those eyes so unnaturally blue they went straight through you and came out the other side knowing more than when they went in.

It must have been something in those eyes and that loping walk and his unnaturally optimistic nature on top of the fact that they put him in those special ed classes that got the Kmart cowboy kids going. "Re-*tard*," they called him everywhere he went. And then they started on the bus every morning, "Re-*tard*!" and Colton just sat looking through his schoolmates with his cornflower blue eyes, forgiving as Jesus, like he truly couldn't feel the pain. It was as if the wiring for pain was faulty in Colton, like he could keep loading up on it forever and ever without ever shorting that fuse. "I don't mind, so it don't matter," he said over and over. By the time he was fifteen, he had learned to swallow pain so deep, you'd never know he could feel it—any kind of pain all sucked up inside of him. But Merinda, eighteen months younger than Colton, couldn't take it. She stood up one day on the school bus and cold-cocked the worst offender. "Leave my brother alone, you little piece of crap," she said, and was suspended off the bus for a week.

. . .

Now Colton kicked a log onto the fire and threw a match onto a handful of dead grass, letting the little flames search around for a grip on the wood. He sank down on his haunches and blew into the smoke, coaxing the warmth into life, and when the fire caught good and strong he put on a pot of water to boil and tucked a wad of chew into his lip. There was woodsmoke in his face, the promise of coffee and bacon and a good day riding ahead of him. He shut his eyes and grinned.

Then Bill came out of the morning and stretched out his hands to the fire.

"Mornin', Dad," Colton said.

"Son."

"Everything okay?"

"Pretty good," said Bill, spitting.

Colton said, "Oh no."

"Yep," said Bill, tugging his hat over his eyes.

Colton was running even before he got out of a crouch, lifting himself off the ground with his fingers, scrabbling for purchase on the flinty soil, like a skater taking a tight corner.

Two horses stood against the trailer, heads hanging, lips loose, eyes shut, ears back and tails swinging against the season's first suck of mosquitoes that had floated in on that morning's early sun. They lifted their heads when they heard Colton coming and Bill's gelding snickered a greeting. "Cocoa!" Colton shouted, but where Cocoa had been there was everything but the mare—a halter, a rope, a rubber bucket. She'd seemingly spirited herself away from the trailer and into the great skirts of the mountains that reached from here to as far as man could hope to walk in a week. Colton spun around on his heels a couple of times. "Crap!" he said. He snatched his baseball hat off his head and flung it onto the ground. "Oh holy crap!"

And then Jake was there, half dressed, pulling jeans over his long underwear. "Colton?"

"Cocoa ran away," said Colton. "Stupid horse! She ran away!"

The boys took Bill's rig and started driving away from the lake toward the heart of the mountains on rough tracks that dead-ended in someone's old campfire or on old logging trails or against the edge of impenetrable deadfall. "You think she'd head home?" Jake asked.

"I think she'd run into the mountains."

"Why?"

"Oh heck, I don't know," said Colton. "She can run wild for all I care." He wiped a thumb down his cheek. "Stupid horse," he said. "If I saw her now, I'd shoot her."

All day they drove back and forth over those ridges, following her trail easily near water, but then her spoor dried up on the rocky soil high up. Colton got out of the pickup and walked slowly ahead, bent over, as if trying to smell the ground. He stopped once in a while to shout, "Cocoa!"

"She ain't gonna come if you call her," said Jake. "She ain't a dog."

"I know that. She's a crappy horse, is what she is," said Colton. "It's getting dark."

"Some bear's gonna get her, or a lion. There's no way . . ." Colton looked around. "Or she's gonna starve to death. There's nothing going on up here. What's she gonna do?"

"She lived wild till you got her," Jake said.

"I know, but that was two years ago. She won't be able to handle it out here."

"She'll handle herself fine," said Jake.

Colton said, "You don't know her." He turned his back on Jake and scrubbed his knuckles under his eyes. "She's very soft. You don't know how soft she's grown."

"By the looks of things," said Jake, "she ain't the only one."

13

BILL'S PHILOSOPHY
OF HUNTING

Bill Bryant had made it clear to Preston and Colton almost to this extreme: if you shoot a skunk, he'd better find you eating skunk steak for the better part of the next week and wearing a skunk hat all winter. If you shoot a goose, you'd better be eating the whole goose, not just the parts most people would say are edible. If you shoot a jackrabbit you'd better be up for rabbit stew and rabbit-skin carpets and rabbit-foot key rings. If you cut a tree for firewood, you'd better check its pulse and make sure it's a dead tree *before* you get your chain saw into its bark. Bill has the utmost respect for anything that can make an honest living in this climate, in part because you can pretty much count on drought, wind, and women to take care of early, accidental death without carelessly contributing to the toll yourself.

Bill Bryant also made it clear to his boys that if you brought something into this hard, short-summered, scarcely covered world, or if you were lucky enough to be put in charge of land or a hunting permit, it was yours to take double care of. This wasn't fat city like California or New York where some welfare group was going to come along and rescue your responsibilities if you didn't take care of them. That went for kids and cats and horses just as much as it went for soil and wives and wildlife. And if you fail to secure your horse properly and she gets away in the middle of the night, she's yours for the rest of your life to find. Not that Bill ever said this explicitly, but since he is the kind of man who can say more in an hour of silence than most men can say in a year of talking nonstop, he didn't actually *need* to say anything explicitly. It just was. And Cocoa being a wild horse, like that, made her even more of a responsibility than an ordinary horse. But Bill didn't say anything, not even, "I told you so."

He didn't need to.

14

LOOKING FOR COCOA

So it would happen that they'd be sitting watching television or cleaning guns and Colton would suddenly say something like, "Maybe she's out on the Hogsback. We haven't looked out there yet."

And they'd pile into Merinda's blue Ford Escort and drive out of town with halter ropes and head collars and .22s and fishing rods and whatever else it occurred to them to grab and someone might ask just before the door slammed, "Where you boys going?"

And they'd shout back, "Looking for Cocoa."

So, two boys in September, barreling down a road that has never existed on any map, burning hours down watching for the shape of a red desert-colored mare against the shapelessness of the high plains and finding nothing in the silences between the spaces. There are tins of Copenhagen on the dash, a stash of Mountain Dew in back, Neil Diamond or Dolly Parton or Kenny Rogers on the sound system, jumping every time the car hits a pothole. To the ordinary eye, there's not much to look at; the road ahead through sagebrush, a couple of pronghorn ante-lope Zen-mastering into a sky so big that this could go on for some time.

But then Colton says, "Think I could catch one of those prongies?"

"If that's what you have to do," says Jake.

"Jake brake!" shouts Colton.

So Jake pulls on the emergency brake and the blue Ford Escort with the little hula girl on the dash disappears sideways into its own cloud of desert pink. Colton unfolds himself out of the car, "Whee-haw!" and begins loping up the road. The antelope lift their heads and watch him placidly until he's abreast with them, and then they run like water uphill and spill into the sun-dancing horizon. Colton bounds up the ridge after them, over old-growth sagebrush, knees coming up to his waist, arms beating time with eternity.

Jake sighs, leans over and pulls Colton's door shut. "Crazy freakin' sonofa . . ." he says, ducking to see through the windshield. Colton's already made it up to the top of the world and now he's just an apostrophe against the sky. The antelope are long gone over the next ridge. The vastness of it all is dazzling and slow, there's no way of catching up with them or of covering a space this endless so that if you were paying that kind of attention, this would be a heartbreaking world.

By the time Colton gets back into the car Jake's almost asleep against the window on the passenger side, cap pulled down over his eyes. Jake says, "Happy now?"

"Born happy," says Colton. He winds down the window, spits a stream of tobacco into the empty world, turns on the car, and sears a couple of doughnuts into the gravel road. A smell of hot sage and fresh dust and decades-old manure fills the air.

"Oh crap," yells Jake, coughing and waving his hand in front of his face, "would you stop driving like an idiot?"

Colton yells back, "I'm not driving like an idiot."

"You are. I'm getting carsick over here, with you spinning doughnuts. Wind up your freakin' window."

"I like fresh air."

"That ain't fresh air, you retard. That's dust you just put all over me. I don't like dust."

"That's because you're a pussy."

"And you're a retard."

"I ain't a retard."

"Well, if you ain't a retard, you sure drive like one."

And then Colton slams his fist across the seat and thumps Jake in the chest and Jake thumps Colton back and then Colton steps on the brakes and the hula girl on the dash is choked in a fresh layer of dust. After that, nothing happens for a few moments. Then Colton says, "I ain't moving this car another inch until we both say we're sorry for what we just said."

There is silence for quite some time inside the car. Outside, grasshoppers crackle hotly on the papery grass and the sun loops threads of heat into the ground. More antelope appear on the horizon. A tiny cloud pokes across the plains.

Colton says, "I'm sorry. I don't really think you're a pussy, Jake."

And Jake says, "I'm sorry, Colton, I don't really think you're a retard."

"Okay then," says Colton.

"Okay," says Jake.

And the boys drive on in silence for another fair bit until Colton suddenly says, "You think I coulda catched one?"

"Catched one what?"

"A prongie."

"Nope."

"But I came close."

"No you didn't."

"Close for me, I did."

"Yeah but," says Jake, "that's not saying much for a retard."

Colton laughs, "He-he-he."

Then they drive some more until Colton says, "You want to know what?"

Jake doesn't say anything.

Colton insists, "Seriously. Do you want to know what?"

So Jake says, "Sure."

"I've decided I've got two ambitions left in this life."

"Oh yeah?"

"Yeah," says Colton. "Number one is, I've gotta find me a good woman."

Jake waits. He waits a little longer. Then he asks, "That it?"

"That's it."

"What's your other one?"

"Other one what?"

"Other ambition. You said you had two ambitions in this life."

"Oh," says Colton. "Holy cow, I forget now. He-he-he."

15

FIREWOOD

The ax came down over Colton's head, wood fell apart. He picked up the log, set it on end, brought the ax down again. Over and over. It was like listening to the repetitive work of a machine, something inured to boredom. They say Colton didn't so much chop firewood as he made toothpicks. Sweat started to trickle down his face so he took his hat off and stuffed it into his pocket, then he opened his coveralls to let the cold in and you could see part of the words on his favorite T-shirt, i put ketchup on my ketchup.

Colton split firewood like the ax was language for whatever he didn't have the vocabulary to say and all the time he was chopping wood he was thinking of a girl. And less specifically, he was thinking of a horse. And although he wasn't thinking it exactly in this way, he was wondering what it would be to possess a girl the way you could possess a horse. Not in the crude, manhandling way, but in the wordless miraculous way, where there was no end to either of you and the possibilities of you, together, were more than double of what they were of you, apart.

Just the other day, he'd asked Merinda, "What does love feel like?"

And she'd said, "Colt, if you gotta ask, you ain't there yet." And anyway, she'd asked him, what kind of boy wants to know about love before they want to know about . . . about, you know?

And Colton had said, "It ain't right to talk about *that*," shocked. "We don't talk about that."

"Holy cow, Colton," said Merinda.

And that was when Colton's rhythm came to pieces and the next time he brought the ax down it cleaved his left foot in quarter, right through the boot, "Sonofa!" He waited for something more to happen because it seemed there should be more to it than this. You chop the side of your left foot through to the bone and that's it? There's not as much as you'd think to a person, Colton was thinking. And then a hot chug of blood started to gush out of the side of the boot and to pool on the ground so Colton felt reassured. "That's alright then." No pain that he couldn't handle and blood doing what blood, in the circumstances, is supposed to do. Colton picked the ax up again, leveled a log on top of the block, brought the ax down, the log split. He set one half upright again, his rhythm back.

He tells himself, "The trouble with love is you get careless and lose your rhythm, that's the trouble with love. You forget where your feet are."

Then he goes inside and takes off his boot and says, "Sonofa!" and puts his boot back on in a hurry. Then he limps across the field, under one barbed wire fence, across another field and under another barbed wire fence, to the neighbor's house and knocks on the door and when a man standing six-foot six in his socks comes to the door Colton ducks and weaves a little, hands shoved into the pockets of his Carhartts.

"Hi Stretch," he says.

Stretch grins, "Well hello there, Colton," he says.

"I was just wondering if you'd mind taking a look at something here." Colton points to his foot and the black pool gathering below his jeans. "I cut my foot a little bit."

And Stretch says, "Dang, Colton. You better come inside."

"I can't do that."

"Why not?"

Colton says, "I'll make a mess on your carpet and I don't think your wife will be very happy with that."

Stretch looks over Colton's shoulder. "You walked all the way from your house like this? Where's your Mom?"

"She's gone to Salt Lake."

"Where are Preston and Tabby?"

"Bowlin'."

"You gonna bleed to death, boy."

"That would most certainly be an inconvenience," says Colton.

"Get in the truck, we're going to the hospital," says Stretch.

"I can't do that," says Colton.

"Holy cow, Colt. You get in that truck."

Colton shook his head. "I ain't gonna do it."

. . .

So Stretch calls Preston and Tabby and tells them that Colton is in the process of bleeding to death on his front porch and could they please come and take him away. Preston drives like a cat with its tail on fire all through town with Tabby shouting, "Slow down!" and "Hurry up!" in about equal measure until they get to Stretch's house. Stretch and Colton are still on the porch in an increasing pool of Colton's blood.

"You better hurry," says Stretch, "he's been standing here bleedin' for about half an hour. Refused to go with me, refused to go in the house. Twice as stubborn as a mule and about half as intelligent."

Preston tells Colton, "Get in the truck."

Colton hops to the truck and Tabby helps pull him into the passenger seat.

Stretch hangs through the window and tells Colton, "Cowboy up, cupcake. No crying at the hospital now."

Preston slams the door and pulls himself into the driver's seat.

"You idiot!" Tabby tells Colton as they burn out back onto the main road, "you coulda bled to death. Why didn't you let Stretch drive you to the hospital?"

Colton shrugs.

"What were you thinking?" says Tabby.

"I figured I didn't want to get blood all over his truck."

"Oh Lordy," says Tabby.

"It don't wash out of carpet so easy," says Colton.

• • •

Which is how Colton ends up in the emergency room where the staff treat him in a way that makes everyone seem like family in Wyoming. "What is it this time, Colton?" says the nurse.

Colton says, "I was distracted."

"Distracted?"

"I was chopping wood and I got to thinking about love."

"Love? My goodness, Colton, you about chopped your foot clean off."

"It won't happen again."

"How many stitches have we put in you over the years?"

"I don't know."

"Well, it's twice as many as any other kid in this town."

"Yes, ma'am."

"Okay." The nurse had the bottle of antiseptic in one hand and a scrubbing brush in the other. "This is gonna sting a little bit but we got to get 'er clean before the doctor puts stitches in. You want painkillers or you gonna cowboy up?"

"I'll cowboy up," says Colton, trying not to let tears squeeze out of the corners of his eyes.

"Good boy," says the nurse and bends over his foot.

"Sonofabitch!" thinks Colton loudly.

"Are we doin' okay?" says the nurse.

"Yes thank you, ma'am," says Colton.

"Colton?"

"Yes, ma'am."

"Next time you get to thinking about love can you make sure you're standing well away from anything sharp or hot or twenty-six feet high?"

"Yes ma'am."

"Good boy. Okay now, here comes another sting."

"Sonofa-freakin'-bitch!" Colton thinks.

16

COCOA

June

———— ◆ ————

A week to the year that Cocoa ran away, Kaylee got a phone call from the Bureau of Land Management regional office to say that a mare with a BLM brand signed over to Bill and Kaylee Bryant had fetched up near Freedom, Wyoming. A horse trader had tried to push her into an auction with a few of his yearlings. So Kaylee phoned the horse trader and he said he didn't know how the mare had ended up in his herd. He swore up and down he'd never have tried to sell her if he'd known she belonged to someone else, and so help him, he must of not counted so good when he was loading his herd up off the range last fall and he hadn't noticed her then and by the time he got to noticing her he coulda sworn she was just like a mare he used to have, so maybe he got a little confused. "Straight up," he said, "I coulda swore she was one of mine."

"Well, we'd sure like her back."

The horse trader thought for a moment and said, "Then you gotta pay me for her keep over the winter."

• • •

Colton was working a double at the burger joint that day, so Kaylee drove up alone to Freedom with the horse trailer the next afternoon. When she reached the horse trader's house, there was no one around. Kaylee knocked on the door, rang the bell, peered through the kitchen window, but the horse trader had obviously thought the better of the merits of meeting with the owners of the desert-colored mare. Kaylee walked around the back of the house. A knot of thin cats untangled off a woodpile and streaked into the dark corners of a barn. And then Kaylee heard Cocoa's husky call. She was standing at the rails in a field near the barn and she had gone stiff and alert with recognition.

Kaylee laughed. "My goodness, Cocoa Bean," she said. "There you are."

She came up to the fence and the mare stretched her neck toward her, breathing hard, her nostrils wide, for a waft of familiar scent. "You're looking fat enough," said Kaylee. She dug in her pocket for slices of apple and fed them to Cocoa over the fence. "They didn't starve you, none, huh? My angel." The horse picked the slices of apple up and crunched them wetly. "Colton was looking for you everywhere," said Kaylee. "You just about had him giving up on you." She ducked under the fence and put a halter on the mare. Cocoa stood solid while Kaylee buckled it up for the first and only time in her life. "That makes a change," said Kaylee, stroking the mare's neck. "Ready to come home now?"

Then she loaded the horse on the trailer and left a hundred and fifty dollars under a stone on the front doorstep for the horse trader's trouble, water, and feed and drove home. Colton was waiting on a fence by the house when she drove up. He followed the trailer around as Kaylee pulled it into the yard and was unlatching the ramp before she had even switched off the engine.

Cocoa backed into the yard and Colton took the end of her halter rope. "Stupid horse," he said.

"You're more alike than you know," said Kaylee.

"Eh?" said Colton.

"Scaring a body half to death," said Kaylee.

17

GRADUATION

———◆———

Jake, Cody, and Colton went cliff-diving in Utah to celebrate the unlikely miracle of their having made it through high school with a certificate to prove it—Colton with a B minus average, helped along by an A plus in Hunter Safety. They drove Merinda's Escort out of Evanston singing Neil Diamond loud enough to ruin paint and Colton so full of himself he would turn down the music every so often to say, "Next thing I'm gonna do is be a rocket scientist," or "I think I'll try a little brain surgery next," or "Those fools didn't find a cure for cancer yet, did they?"

Until Jake said, "Easy there, Colt. It was just high school."

"I know," said Colton, but his knees were hopping up and down. "Who'd ever have thought it? I'm a freakin' *genius*."

"I guess by your standards," said Cody.

"I'd much rather be Reverend Blue Jeans," sang Colton.

"Holy cow, Colt, you don't even know the words to Neil Diamond by now, how you gonna fly a freakin' rocket?" said Jake.

"It's, 'I'd much rather be forever in blue jeans,'" said Cody.

"He-he-he," said Colton. "I like it better my way."

Then they listened to the Nitty Gritty Dirt Band for a while and the land got redder and harder and the trees shrank into plump green cones no taller than a man and there didn't seem to be much for a cow to eat. But it was beautiful and otherworldly and you could see how a place like this might inspire fanatical, half-crazy religion in a person. It smelled different too, without

Wyoming sage all minty sweet on the air. Now the air smelled of sun and rock and ancient ritual.

"I don't think I'll ever leave Wyoming," said Colton, staring out of the window.

"Why not?" said Cody.

"I dunno," said Colton, frowning, "it just seems as if Wyoming likes me."

"Don't be a retard," said Cody. "What are you now? A freakin' poet?"

"Eh?"

"'Wyoming likes me.' Now that sounds exactly like a whiny-assed, poet-fairy thing to say."

"He-he-he," said Colton.

· · ·

When they got to the cliff, Colton was first out the car. "Here we go, boys! Time to swim!" He had his shirt off and was swinging it around his head. "Whee-haw!"

"Holy crap," said Cody.

"You can't swim," shouted Jake out of the window.

"He *still* can't swim?" asked Cody.

"Not so as you'd notice."

"Holy crap."

Colton stripped off his shorts next, limping and hopping, tangled in his belt, and then he stood on the edge of the cliff with nothing on but his boxers. The water, twenty-six feet below him, was running fresh off the mountains, snowmelt all pale green and dense with cold. Colton's flesh stood up in prickles of goose-bumps. He spun his arms around and around, backwards and forwards until they were fit to wrench right out of their sockets. The fly of his boxers was gaping open, although you could tell he hadn't noticed, the way he was grinning like a fool and everything was hanging out for all the world to see. There were girls on the edge of the cliff too—they had just arrived in a Chevrolet—and they were propping each other up, giggling, bundled up in sweaters and scarves.

"Holy crap," said Jake, "his Old Glory's hanging out."

"Holy crap," said Cody.

Jake got out of the car, followed by Cody. "Hey," he yelled, waving his hands, "jump!"

"Come on, you pussy," yelled Cody, "jump." He was out of breath. "Holy crap," he said to Jake, "he is completely clueless."

"What about you?" shouted Colton back.

"Jump!" shouted Jake.

"I just got to get me warmed up a bit," said Colton. He started to jog in place and run in circles, his knees pumping, his arms going like helicopter blades.

The girls folded up over themselves and squealed.

"Hey, look at that," said Colton, pausing his warming-up exercises. He pointed to the car. "Those girls are definitely checking me out."

"It's because your Old Glory is hanging out for everyone to see," said Jake, breathing hard with the effort of running.

"That's what we been trying to tell you," said Cody, arriving at the edge of the cliff. "Your fly is wide open."

"Well if they ain't seen one yet, there's no better time than the present," said Colton, spinning a few final laps, waving and smiling at the girls like someone who just won a contest. Then he jogged to the edge of the cliff, spread his arms out, let go of the earth, and hollered like Tarzan, "Ahhhh-ah-ah-ah-ahhhh," his legs wrapping around the air like corkscrews. When Colton hit the water, Jake peered over the edge of the cliff to see him struggling in his barely drowning dog-paddle for the bank, chin lifted to the sky. "I did it! I did it!"

The girls laughed and clapped and said what an idiot that boy must be to jump so high into a river and what a double idiot to swim in this cold water and what a triple idiot not to know his fly was open. After that they got chatting to Jake and Cody and someone suggested building a fire, which seemed like a good idea to Colton who had come shivering up from the river by then and before long a great crackling bonfire warmed the edge of the gorge and someone brought out marshmallows and someone else had thrown scarves over Colton's shoulders and around his neck and everyone was talking about how Colton had been standing there on the edge of the cliff with his Old Glory out for everyone to see and Colton was laughing, "He-he-he," and jogging in place to get warm, dressed now in nothing but wet boxers and a wrapping of scarves. "He-he-he."

18

BULL RIDING

All Over the West

———◆———

A few days later, the rest of their lives started.

Jake knew he'd end up on the oil patch sooner or later, but meanwhile he got a job at the state mental hospital in Evanston working with the kids for a year. He was good at it too, maybe because he'd come close enough to landing a spot in the bin himself and he understood how easy it is to misplace yourself out here: loaded guns in unlocked closets, crack-cooks in trailer parks, windows painted black to block out the sun in the high lonely hell of the middle of nowhere. Vicious circles, Jake knew, could feel particularly vicious when they were closing in around your own neck.

Cody got on the bull-riding circuit. He got lined up at small rodeos in some towns you've never heard of and some towns you've not only never heard of but that you couldn't even find on a map. And still that boy rode and rode as if he were in the big time, the uncrowned bull-riding King of Las Vegas, the Emperor of Denver, the Terror of Cheyenne. No matter how many bulls dislodged him, or got after him, or nearly gored him, Cody never seemed to get so that he wanted to quit. But neither did he seem to get any better at it. He rode all the big sky a small-time rodeo bucking bull could throw at a kid and he still couldn't feel his way to the end of eight seconds.

Colton emptied the bank of all the money he had, which

wasn't much, and he chased rodeos with Cody. The way Colton saw it, a boy has to do what a boy has to do, and if it was riding bulls, then so be it, and a boy riding bulls in little wiry western towns needed someone to cheer him on. So Colton drove the Mazda to the rodeos and pulled the bull-strap once they got there and stood with a boot propped up on the rails of whatever no-name rodeo arena he found himself at and he yelled encouragement at Cody and he slapped Cody on the back when it was all over, put ice on the bruises, and kept the Mountain Dew cold. So you could never say that Cody and Colton didn't put a lot of heart into bulls that summer. And in between the kicked ribs, and the long hours on the road, and the crappy food, and the nights sleeping on saddle blankets in some rancher's field or in warehouse parking lots on the outskirts of town—and a lot of times even then—they had a whole bunch of fun.

"This time," said Colton, driving into the dusty parking lot outside Pinedale in the Upper Green River Valley for the mid-July rodeo, "keep your chin up and look at the horizon."

"What in heck do you know about it?" asked Cody, spitting.

"What my dad's told me," said Colton.

"Exactly. You ain't never done it."

"Sure I did. I've rode bulls a time or two," said Colton. "Wherever you look is where you'll end up."

"Well then, you must've been looking up your ass."

Colton laughed, "He-he-he."

"Anyway, I don't want to end up on the freakin' horizon," said Cody.

"Well, it's better'n ending up on the freakin' ground," said Colton.

"Okay, okay. Just shut up already and let's get 'er done."

So Cody put on his vest and his leather glove and he taped his arm and worked the cricks out of his ribs and his neck and strapped on a knee guard while Colton checked the roster.

"You're in chute two."

"Which bull?"

"I forget," lied Colton.

"Holy crap, they're gonna put me on the Ripper, ain't they?" said Cody.

. . .

On the other side of the arena, there were maybe fifty people on sun-buckled bleachers, sharing picnics and blankets and passing bundles of babies between them—a community made intimate by the public business of watching their sons and lovers and other people's sons and lovers get thrown around like rag dolls on broncos or bulls. On tailgates on either side of the bleachers, people had set up picnics and folding chairs. A few cowboys were already a little drunk in that optimistic, shiny-eyed way you get from drinking weak beer on a hot day.

Then a little girl in a white cowboy hat galloped into the arena with a Stars and Stripes bigger than the body of her horse and belted out the national anthem in a high, brave voice and Colton and Cody and everyone else with a heart covered it with a cowboy hat. A horse tied to the trailers called out for its herd and the little girl's horse stretched out his neck and cried back and just like at all rodeos from one end of the country to the other, America was born again in all its sentimental, painful bravado. Then the music ended and everybody with a hat jammed it back on their heads and the little girl galloped out of the arena and the serious business of the rodeo began.

. . .

The first bull rider was released into the arena, his fist in something like a Black Power salute, spurs stroking back and forth along the bull's shoulders. Eight seconds is as long as it takes to read this sentence, and the whole world contained in the violent dance of those moments. But the cowboy didn't make it to the eight-second bell. He was thrown down hard about three hard, screwdriving twists into his ride. Even from across the arena you could hear the wind leave his body, the insult of flesh connecting to earth. The bull continued to buck and twist, spitting sand up against the rails. Then he registered that his rider was no longer on top of him and so he flattened out and spun around the arena a couple more times looking for the offending party while the clowns tumbled around and tried to distract him with flags and hollering. But the bull didn't care about the clowns. He had a look in his eye that suggested he had some unfinished business with mankind, cowboys in particular. Eventually the pickup men managed to get behind the angry beast with horses and chase him out of the gate. The bull still looked murderous.

"Holy crap," said Cody.

"Just do your freakin' stretches," said Colton.

Cody uncertainly caught his ankle behind his back, held on to the rail in front of him, and tugged. Since the cowboy who had just been thrown by the bull didn't look as if he was getting up anytime soon, the commentator started up with a stash of "my wife" jokes. An ambulance edged between the row of jeans and plaid shirts that lined the fence, but the commentator sent it back. "I don't think we need the meat wagon," he said. "He's still alive. I saw his toe twitch."

More of nothing happened. The sun was setting now behind the great, glacier-sweet Wind River Mountains. "Look at that, folks," said the commentator, "a real Wyoming sunset." And it was, if Wyoming means big sky bleeding into mountain peaks, mad bulls, and a flattened cowboy. At last the cowboy stirred and was helped to his feet. He hobbled out of the arena.

"No score for that cowboy," said the commentator. "Come on, boys, let's *rodeo*."

Now Colton hung over the chute with Cody white-faced and hardly breathing, one leg over the top rail, ready to lower himself onto the bull contained in the tight area below.

"Breathe, Cody, for the love of crap. You're gonna pass out if you don't breathe."

Cody gulped. Then he nodded, settled himself on the salty black back in front of him, and Colton tightened the bull rope around the bull's flanks and someone else opened the gate and there was a pile of dust and then it cleared enough and there was Cody getting slapped up and down on the back of the bull and every time that slab of bull came up to meet Cody's back you could hear the air leave the lungs of *both* of them, "Uh-huh," the deep grunt of the bull and Cody's air coming out in high painful exclamations. It sure makes a body wonder, doesn't it, that if cowboys are such a myth, then how do you explain that their pain is so real?

"Ladies and gentlemen," said the commentator, "Cody Eaton, all the way up from Evanston, Wyoming."

There was a splattering of applause.

"Hold on, Cody!" Colton yelled. "Look up! Look up! Look at the freakin' mountains. Oh, holy crap."

"Riding the bull we like to call the Ripper," said the commentator. "Ladies and gentlemen—this animal is one ton of hide, hooves, horn, and hate."

"Holy crap," said Colton again, watching Cody get whip-lashed.

And then there came the "O" expression on Cody's face as he ran out of seat and came hopping up the bull's neck, ran out of neck, and then hit the dirt with his face first. Now he still had to jump up, with what few wits he had left, and run for his life while the bull swung his head around trying to get the boy in his sights. The clowns ran into the arena then and the country music went up, uncomplaining and nostalgic, and Colton scrambled around to the gate to open it for Cody.

"Man, you almost had it," Colton said.

Cody had a reverse-raccoon face, mud where sand had combined with sweat framing white, frightened eyes. "Seriously?"

"Seriously. For a moment there, I thought you were definitely gonna walk out with the belt buckle."

"Really? You think so?"

"I *know* so. They just put you on the sonofabitchest bull they got. Man, they *robbed* you."

19

PARADISE ROAD

Upper Green River Valley

It was dark by the time Cody and Colton left the rodeo grounds. The team-ropers were still left to ride—midnight cowboys who still had to go home after riding and feed their horses and check on cows and pray to God for better rain and do whatever else it is cowboys do before rising again at dawn to do it all again. Hardly anyone stayed to watch the rest of the rodeo once the bull riders had gone. Behind Cody's Mazda, as they drove away from the arena, a trail of pickups and horse trailers reflected pink dust in their brake lights, and behind the line of cars and the commentator's box, the Wind River Mountains sheared blackly against a moonlit sky. In front of the boys, beyond Highway 191 onto which they soon drove, were the great high plains of Wyoming, a seemingly endless dark swell of silence, like a ship-less sea.

"Let's sleep out there," says Colton, pointing out into the plains. "They don't charge you for your money in God's hotel."

"I heard about some idiot got lost out there a few summers ago and the cops found him with his head in a badger hole."

"Doing what?"

"You never heard about that?"

"Nope."

"I'm telling you the honest truth."

"Was he looking for badgers?"

"I don't know about that. By the time they found him, he was dead."

"Holy crap."

"Yeah. Fried his ass in the sun from what I heard tell."

"Holy crap," said Colton again.

"Most likely a meth head," said Cody.

"No kidding."

"That's a losing war, for sure," said Cody.

"Most definitely," said Colton.

The boys found a gravel road that led out onto the plains, Paradise Road it said. The road followed the New Fork River's descent from the mountains into the rest of the great wide world. The little Mazda's headlights picked up wave after wave of sagebrush as far as the imagination could stretch, the swell of blue-green replenishing itself, always more where the last mile came from. Taken one step at a time, a person might think never to eat all the miles that make up these plains, just as it might seem impossible to swim the sea one stroke at a time. Although, like the sea, the endlessness of the plains' open abundance is an illusion. Nearly all the plains are already swallowed up, paved over, plowed under, flattened, hardened, drilled. You can drive across what's left of what is wild in an afternoon or less.

Colton wound down the window and stuck his head out. "Smell that! Whee-haw!"

Cody hung out of his side and the boys drove along like a couple of hunting dogs, breathing in the summer-cool mountain air and the salty scent of deer and antelope and the vague hint of a wildfire somewhere west of here. Once in a while a pronghorn startled out of the sage and sped in front of the pickup before zigzagging back into the darkness. Two or three lonesome ranches pooled light from the top of barns or from the front of a porch. And then, right far out in the distance, like a couple of tiny Eiffel Towers lit up for the sheer romance of it, there were two drilling rigs powering into the earth after pockets of natural gas.

The boys found a flat spot near the road behind a sign that said this land was administered by the Bureau of Land Management and that it was critical winter habitat for big game, closed to off-road traffic in the summer and closed to *any* traffic during the winter. "They sound kinda paranoid about this place," said Colton. He made a circle of stones and looked around the sagebrush for firewood. "Man, ain't nothing much out here to burn," he said. "I'll have to break out our emergency store." He dug around in the back of the Mazda for wood, then he made a fire. The boys laid down a couple of sleeping bags upwind of the smoke and downwind of the Mazda. Then Colton pulled out a Tupperware of Kaylee's meatloaf and a couple of cans of Mountain Dew and he said, "This is the life, eh?" The boys ate and drank and dipped a little chew and then they stretched out on top of their sleeping bags and tucked their hands behind their heads and watched for slow satellites sliding through the blackness, sending television and weather reports and telephone conversations back to earth, and shooting stars, which, in a sky this dark, showered down too numerous to count.

"How you feelin'?" asked Colton.

"Okay."

"You got pretty beat up by that sonofabitch bull."

"Not too bad," said Cody.

Colton grunted.

"I'll make it to the eight seconds next time."

"Sure you will," said Colton.

"Sure I will," said Cody.

Then the boys went quiet and watched the sky some more.

"Do you ever wish when you see one of them shooting ones?" asked Colton.

"Sometimes," said Cody.

"So what would it be? If you had one wish."

"To be bull-riding champion of the world," said Cody.

"Holy crap," said Colton. "That's a good wish."

"What about you?" said Cody.

"I guess," said Colton. "I wish I could be just like my dad."

"Holy crap," said Cody. "I dunno if shooting stars can do all that. That's gonna require the entire universe to collapse."

"He-he-he," said Colton.

Cody spat. "Ha," he said.

20

DRILLING ON THE RIGS

Utah

So it went all over Wyoming, Utah, and Idaho from one small-town rodeo to the next until Colton ran out of money. "Mind over matter," he said then. "I don't mind, so it don't matter." And he pawned a watch to fill up the little Mazda's tank with gas and off he and Cody went for another round of punishment in another dusty arena in another place where the coffee tasted of unfiltered ditchwater and Stars and Stripes shredded in the dry wind from the tops of the commentators' boxes and America wore its heart in its mouth while bulls ran wild over sand and the sun set like heartbreak. But by the middle of August Colton found himself having to write bad checks to help pay for the Mazda's gas—which didn't feel so bad as it sounded. The way Colton saw it, checks were just some kind of citified promise that if he *had* had the money, or maybe when he got the money some-day, he'd be good for the whole amount—bad checks written with good intentions, in other words.

Still, in a small town like Evanston, it didn't take long for the bailiffs to catch up with Colton and explain to him that good intentions are the way straight to hell, or at least straight to a cell-bed in the city lockup. Kaylee had always told her children, "You land in jail and you better not call me." So Colton sat on his hands for a night but in the end he phoned his mother and of course she bailed him out and Colton was a pawning fool to pay her back—he pawned his DVD *and* his gun collection, his cowboy boots, and half of everything else he owned—and he told Cody that he was very sorry to leave him at a time like this, when Cody was so obviously close to the big buckle, but he had to give up rodeo running and start being a man.

And the day after that Colton signed up for the safety talk and the piss test and he signed the piece of paper saying he'd read the manual about this, that, and whatever else and he went out on the rigs down in Utah where Bill and Preston were drilling and he worked as many hours and as fast as they'd let him to make it up and jumping around like a frog on a hot rock is how he slipped one day a year into it and broke his foot, which is how patterns repeat, kind of, because Colton's grandfather (Bill's father) had broken *his* foot on a rig near Riverton pretty near fifty years earlier, which was how Bill came to be born in a hospital just down the corridor from where his father lay with his foot in a cast. Except back in those days the drilling companies didn't have any kind of safety policy to speak of, so as soon as he could walk again, old man Bryant was back on the rigs.

Whereas Colton . . . Well, he wasn't fired exactly, but the drilling company was getting pretty paranoid about their accident record in these careful days and boys falling off rigs weren't how they like to report things. And Colton, like his father and his brother and every other roughneck on these patches, was nothing more than a part-time laborer, so once his foot was healed he was a no-time laborer. Jake was in the Upper Green River Valley by then, working on the new gas fields as a flow tester, so Colton packed up and moved up north to work with Jake almost exactly where he and Cody had been camping that magical night after the Pinedale Rodeo when Colton had wished on star after falling star that he would grow up to be exactly like his father.

21

ANATOMY OF AN OIL PATCH

Upper Green River Valley

It's hard to explain how many different ways of being on an oil patch there are because until you've worked out on the patch for a while, the various positions sound unintelligible, like the navigational waters on the sea might sound to someone who has never been a sailor—drilling location NWSE Section 35, Township 32 N, Range 109W, for example, would mean that a rig was about twenty miles southeast of Pinedale, Wyoming, but how would you know that without learning it in your bones and blood?

And the names of the various parts of the rigs are like a ship too, mysterious to the layperson—stabbing basket, draw works, beaver slide, drilling floor, v-door, cellar. And the rigs are manned like ships, in back-to-back twelve-hour shifts by roughnecks, five or six per crew, serious in boiler suits and military-looking boots and hard hats and safety glasses as if the thing might sink unless its drill is being driven relentlessly into the ground.

Roughneck positions sound like characters in a deck of tarot cards: the tool pusher, the floor hand, the derrick hand, the driller, the motor hand, the directional driller, the chain hand. These men—mostly men, the very occasional woman—who appear as tiny dots against the great swell of land and the massive intrusion of a rig, look brave and competent and powerfully equipped. And yet, with the big sky thrown around their heads and with such an unwilling earth beneath their feet (you can hear the ground's reluctance to be drilled from a mile off)—even with God and the president of the United States on their sides—it's hard to see how they can possibly win whatever fight it is they've taken on.

And beyond the rigs there are roustabouts, who service the rigs and wells and keep the oil field running. And beyond the oil patch itself are the CEOs and managers, PR men and politicians, who negotiate and strategize in offices and back rooms far from the noise and rotten-egg smell of the real work. And all the time, the inconvenient biology of human bodies creating logistical and law enforcement challenges for the communities that host the oil-field workers—food and porta-potties, beds and trailers, drugs and sex—because the humans involved in the process of oil drilling aren't always robotic extensions of their drill bits.

Since it's nothing new, there is a name for the depression and lawlessness that comes to communities that have been blessed with the dubious gift of nearby mineral wealth: Gillette syndrome. The psychologist ElDean Kohrs coined the term in the 1970s and popularized it in a paper entitled "Social Consequences of Boom Growth in Wyoming," in which he describes the ills then being visited upon the coal town of Gillette, Wyoming. A boomtown, he explained, experiences an increase in crime, drug use, alcoholism, violence, and cost of living, and a decrease in just about everything good except, arguably, money.

But identifying your history doesn't stop you from repeating it, so if you were to look at a map of Wyoming, circa 2007, and if that map showed all the mineral leases in the state as a red dot, the map would show up as if drenched in blood. And to stanch the flow of carelessness, they have had to build a new fifty-bed jail, a new courthouse, a new drug rehab center in the Upper Green River Valley. Who else to cater to all the new methamphetamine addicts, speeding to stay on top of twelve-, eighteen-, twenty-hour shifts, or speeding to keep the stillness of the high plains from stalling them to death? And *still* you can't police the panic of a boom from rotting the soul out of a place because somewhere in the throes of an energy boom isn't so different from a person in the throes of addiction: there's the denial that things are out of control; there's the sleeplessness and moral carelessness, and the fact that you know that you're doing something that isn't good for you, but you just can't stop.

Oil company representatives attend oil and gas lease auctions in Cheyenne once every two or three months. The successful bidder, the oil company (Ultra Petroleum, say, or Exxon), then hires a drilling company (Nabors or Patterson-UTI, for example) to drill a well fifteen thousand feet into the ground. After that the oil companies hire hydraulic fracturing crews (Halliburton or Schlumberger) to detonate explosions under the ground, which releases the gas, which the oil companies then pipe out of this oil patch to (in this case) California, mostly.

Clusters of metal signs swing in the wind next to gravel road junctions all over the high plains of Wyoming, making a squeaking, lonesome noise, like children's swings in an empty playground. The signs give the name of the drilling company and the rigs' numbers so that roustabouts and hotshotters (people who rush spare parts to rigs so that the drilling never has to stop), ambulance crews or the sheriff, drug testers or porta-potty cleaners can easily find their way to a particular location—Patterson-UTI 455 would mean that a well is being drilled by Patterson-UTI on rig 455.

A single well location scrapes up about seven acres of earth and sagebrush to accommodate a rig, flowback pit, the trailers and vehicles. Then there are compressor stations the size of two or three football fields that condense the gas so that it can be piped out of state, and of course roads must be built across the high plains. And man camps erected to shelter the oil-field workers in their twelve hours away from the rig—rows and rows of trailers with single beds in tiny cubicles lined up across the sage like a kind of high-altitude, open-air prison or army camp.

Taken from the air, this spread of wells across the state translates as if the high plains are experiencing contagious balding, clumps of ground cover falling out and vaporizing. And what is done out here is indelible. You can still see, as if their wheels creaked to the coast only yesterday, where wagons crossed here in the 1870s. And now, on top of the Oregon Trail (sometimes directly on top of it), here is our new history of panic and greed, of loss and carelessness, etched like an accusation for the future to read. One scar over another. Wound upon wound.

Picture an oil patch, then, such as the one created on the high plains of the Upper Green River Valley below the witnessing bulk of the Wind River Mountains and spilling onto the foothills of the Wyoming Range. It is a rolling sea of sage, scraped up and graveled over, humming with machinery and engines, and men and women in hard hats made insignificant by the proximity of the rigs, hundreds of feet high. This oil patch is an accident of politics and war, high fuel prices and Hurricane Katrina's devastation of wells in the Gulf. It's a place of repetitive, machine-powered tedium, a methodical siege interspersed with quick, sometimes fatal violence. Twenty-four hours a day, seven days a week, in blowing sideways ice and snow or the cremating heat of summer, the drilling goes on and on and on spinning through the earth. All through the fall and spring migration of antelope and mule deer, all through the sage grouse mating season, all through the haying season, nonstop through the dead of winter. Steady as a heartbeat. Unstoppable. Mandated.

On and on and on.

22

FLOW TESTING

Upper Green River Valley

<hr/>

Colton is damn near bored to death flow testing because flow testing is what comes once the rig has already drilled the hole and moved on. Flow testing means a lot of sitting around in a trailer on location—twelve, eighteen hours at a time, however long it takes is how long it takes—staring out the window at the over-world of an oil patch while Halliburton or Schlumberger blow up the underworld to get at the gas. And then the frac crews' explosions, which shake the windows in the ranch houses on the edge of the oil field, release a pocket of gas and suddenly it's all fireballs and action stations for the flow testing crews. Colton likes that part of it. He just doesn't like the hours and hours of waiting in between. And most of all he doesn't like flow testing because flow testing isn't drilling. And working on a rig is all Colton wants to do, because that's the only way he can get to be exactly like Bill.

Colton's on the edge of the flare pit now stamping out a few flames that have escaped. The high plains burn dryly, sending up the scent of scorched sage, powerfully minty. In the flare pit, the fire measures the width of a small house and is taller than a man. Colton's a small dancing figure in its glare, like some kind of spirit-man, worshipping. Jake looks at him and Colton's wild, ancient-looking dance. "Holy crap, Colt," Jake says.

"Whee-haw," says Colton leaping on top of the stray fingers of fire. He waggles his hips and spins in a circle and the flames lick his boots.

But he gets the fire put out and the flow testing crew gets the pipe sealed off to harness that particular pocket of gas, and then it's back to waiting again while the frac crew blows up another layer of earth, blasting fine particles of sand deep into the fractures they create to prop open the tiny, tight chambers that hold the gas. And Colton is bored again.

So with his first couple of paychecks, Colton buys an entertainment center the size of a small Radio Shack showroom. The way Jake tells it, they pull up to the location one day and Colton says, "Hold on, boys, just need to get myself situated." And he's got a full-size television with a twenty-seven-inch screen, a home entertainment unit, a luggage bag full of games and DVDs, every version of Game Boy known to man, the full-size steering device to go with it, Nintendo, *Pokémon, Shrek, Toy Story*, you name it. Colton says, "Man I ain't afraid of much, but I'm scared to death of boredom."

Then the flare pit goes up in a fireball and the flow testing crew piles out with Colton up in the lead and he's laughing, "He-he-he."

"What you got now?" says Jake.

"Marshmallows," says Colton, emptying out one pocket and showing Jake half a bag. "For roasting."

"For the love of crap, Colt."

"And a hot dog," says Colton, pulling a bag out of the other pocket.

"Are you freakin' kiddin' me?"

"Don't knock it till you tried it."

"You ain't thinking of roasting those on gas?"

"Yep."

"Man, it's pretty nasty stuff," says Jake, although he says this with the tenderness of a trainer for a promising if unpredictable horse because Jake loves gas as intimately as if it were something living, with a mind of its own. He knows how it behaves underground. He understands gas above ground too, the way it ignites in the flare pit, how it is compressed and piped out of here down to Opal and from there to California. He can tell a high-producing well from one that will disappoint and he knows good gas from the kind that can turn nasty on a person. Before Colton started working with him, Jake was hit once by a faceful of sour gas that knocked him flat on his back and shut out his lights for a few moments and that would have been the end for Jake, except that he was following the buddy system and a coworker pulled him clear of the flume.

He tells Colton now, "This gas can most certainly be horrible."

But Colton is determined. "Come on, it's just fire." He has a hot dog on the end of one fork and a marshmallow on the end of another. "Here," he says to Jake, "have the first bite. How bad can it be? Go on, eat it. I made it for you."

"No, I will not."

Colton says, "Okay then, your loss." He takes a bite. His face changes and he spits. "Sonofa! That's the nastiest-tasting hot dog I've ever had in my life."

Jake starts laughing. "Maybe you need a little ketchup with that."

"I'm not kidding. Here, you try some."

Jake shakes his head. "No way," he says. "I'm fine."

Colton jumps on his friend. "You are not fine. You will have some hot dog." And he has Jake in a headlock and he's squashing the hotdog into Jake's mouth. Jake's spitting and squirming about and Colton's laughing, "He-he-he." And then they're told to stop foolin', there's work to be done and this is serious business.

"Serious business," says Colton, smacking the top of Jake's head.

Jake kicks Colton in the pants. "Retard!"

"Pussy!" says Colton.

23

THE ASTRO LOUNGE

Rock Springs

————— ◆ —————

They'd just come off a twenty-hour shift and there wasn't another well for them to look at for at least a couple of days. Jake had packed the night before and he was out of bed before dawn to drive down to Colorado for a night or two for the only reason any man would ever drive all day after thirty-six hours of almost no sleep—a woman. He was standing in the white light of the refrigerator contemplating his breakfast choices, which were limited to soda and a few hardened Pop-Tarts.

"Hey," said Colton.

"What you doin' up?"

Colton scratched the back of his head and opened a can of Mountain Dew. "I heard you. Just thought I'd better tell you to keep 'er on the road."

"Sure."

Colton cleared his throat. He was twenty-two and although he'd had his heart broken a couple of times, there was no serious evidence to suggest that he was particularly lucky with the women. The trouble with girls and Colton was that they were much more likely to adopt him as a brother than invite him into their beds. He said, "I guess she's pretty."

"Sure she is," said Jake, walking outside.

"The air feels good," said Colton, following him.

"Don't die of boredom," Jake said.

"If I should die before I wake," said Colton.

"Feed Jake," said Jake.

"Are you sure you don't need me to drive down with you?" asked Colton.

"I'm sure."

"Even if I don't drive like a retard?"

"Even then."

"You could sleep while I drive."

"I don't need to sleep."

"Okay." Colton sat down outside the trailer in his pajama bottoms and he watched Jake drive away. "Mind over matter," he said to Jake's brake lights. The sun was firing up across the horizon and the plains were just waking up, mountain bluebirds chasing each other around the sage, a few pronghorns melting away from the highway. The Wind River Mountains stared down at the great wells in the high plains and between the two was a thin slice of soil, where cattlemen fought life on real terms, half crazy for the love of cows and for a tough life. Looking at the slow progress of black cows against the pale green plains made Colton think of moving cattle with Cocoa. And thinking of Cocoa made Colton homesick. He sighed and stuffed a wad of chew into his lower lip.

. . .

Now, at noon, Colton was folded over a video game, bare-chested in a pair of boxer shorts with a mini-fan aimed right into his face. The midsummer heat seemed to be breeding with itself to create baby pockets of heat that crawled under the skin and got behind the eyes. They were predicting a high of ninety today, which was nothing, in the scheme of a national heatwave, but without air-conditioning and with all those bodies piled in next to one another, the rented double-wide trailer felt downright soupy. The television said it was 118 degrees in Phoenix; 117 somewhere in South Dakota. In California, they said, 163 people and 25,000 cattle had died from the heat and chickens and turkeys were cooking in their own skins, 700,000 roasted to death so far. Fifteen pets had died. The governor of California ordered every state-owned fairground to operate as a cooling center. Energy prices went up and up and there were blackouts and brownouts because there wasn't—nor will there ever be—enough natural gas to pipe to California from the whole of the Upper Green River Valley, or in the whole world, to cool that kind of heat.

Colton held a cold can of Mountain Dew on the back of his neck.

The rest of the boys were lined up on the sofa, watching television. "Pinedale has got to be the boringest freakingest place in the entire nation," said one of them. "Nothing to do."

"No one to do it with," complained another.

"No bowling alley," said one.

"No nothin'."

"They don't even have a freakin' stoplight."

"Who wants to go shooting?"

"Shoot what?"

"I dunno. Just shoot. They got prongies."

Colton didn't look up from his game. "Out of season," he said.

"So?"

Now Colton looked up. "You can't be serious."

"Why not?" said one of the boys, winking.

"Out of season? No tag? That's the worst possible thing you can do."

"Worse than murder?"

"It *is* murder," said Colton.

Everyone laughed, except Colton.

"They need a shooting range, is what they need," someone said.

"They *got* a shooting range. What'd ya think a stop sign is?"

Everyone laughed again.

"They need a titty bar is what they *really* need," said someone else.

Then one of the boys sat up and slapped the back of the boy next to him and suggested that they go down to Rock Springs and get an Astro burger.

"You ever had an Astro burger, Colton?" one of the boys asked.

"No," said Colton.

"You ain't *never* had an Astro burger?"

"No," said Colton, "but I'd drive three hundred miles for better food than they serve 'round here."

There was more laughter.

And Colton said, "What? What I say?"

Someone said, "You never heard of the Astro Lounge, Colton?"

"Oh," said Colton. He stuffed a wad of chew into his lip. "Sure," he said. "Sure I heard tell of it."

. . .

The two-hour drive from Pinedale toward Rock Springs bleeds the landscape plainer and plainer, but, to be fair, after the gathering beauty of the Upper Green River Valley, anything would be dull. In the Upper Green, rivers and mountains, high plains, and the remains of the wilderness pile upon the eye until the soul hardly believes it isn't in paradise. They say the untamed magnificence of this valley drove some of the early explorers and trappers wild. Mountain men fled into the foothills and people said they must have been bit and maddened by the glory of it all.

But Rock Springs, as if compensating for the surpassing richness of the Upper Green River Valley, has never been a pretty place to make a living. The future wife of Alexander Graham Bell, Mabel Hubbard, wrote to her husband in 1887 of a trip through the early settlement, "We are stopping now at a coal station. . . . There are coal mines around . . . but the houses are all of the poorest description and look temporary. There is no appearance of home about them as there was in even cold and dreary Laramie. There, the land is cultivable and settlers have made homes. Here, the fetid waters make man's stay one of necessity, never of choice."

And Rock Springs hasn't recovered from being a dirty little town in the decades since then. A glittering, hard boom-and-bust settlement surrounded by beautiful layered pink and white desert dunes, which a *60 Minutes* program in the 1970s contended was still stuck in the Wild West, and not in a good way. Boom *or* bust, there's the breath of disappointment in the air. Like nothing ever turned out the way anyone thought it would, and everyone's worst nightmares have come true. If La Barge or Wamsutter or any of the other little flash-in-the-pan towns in the West are like waking up the morning after the night before with a beer hangover, Rock Springs is like waking up after a weeklong methamphetamine binge. It looks like a town thrown together in the throes of a temporary fit of panic—cheap clapboard trailer parks and blowaway boomtime mansions confined by big-box stores. Even the people who love it, love it the way a parent protectively loves their roughest child—because no one else will.

So the Astro Lounge on Pilot Butte Avenue in Rock Springs isn't much to speak of either, even if you go in for that sort of thing. It's just a dark drinking hole of a strip club—"crusty and unedifying" is the way one ex-bouncer describes it—but a lot of the boys on the oil patch in the Upper Green and all over Rock Springs know it like it's heaven with a side of fries. It smells of stale cigarettes, spilled beer on rank carpet, old sweat, deep fried meals. The bathrooms have a hospital scent to them, antiseptic and vitamin B, which is what the dealers use to cut their meth. But the girls are friendly enough and some of them are downright gorgeous, especially to lonely men used to the rough company of other men and made hungry by Wyoming's notorious lack of women.

· · ·

The boys ducked into the relieving, air-conditioned chill of the dark lounge and showed their driver's licenses to the bouncer. The music was loud and repetitive like it was trying to drown out thought. They found a quiet table near the back and everyone ordered a beer except Colton, who said he'd have a Mountain Dew, please. The waitress was pretty in a desperate, worn-out sort of way with hair so bleached it had deconstructed into a mild frizz around her face and sad green eyes, but she had strong, childrearing arms and long muscled legs. Her voice was husky with smoking. She asked if anyone wanted to have anything to eat and people placed their orders and Colton asked for a double cheeseburger with fries and extra ketchup, please, and he looked up at the waitress in that way that he had, and she sucked in her breath with those eyes looking through her and she plain couldn't help herself. She leaned right over him. "And *anything* else for you, blue eyes?" she asked.

Colton looked down quickly and said, "No thank you, ma'am."

Up on the stage two women were thrusting their bits and pieces at the audience and at one another and their underwear looked unequal to the task. Colton didn't know where to rest his eyes in the wait from now until the food appeared. He sucked down the last of his Mountain Dew and wished there was a window so he could look outside at the sky or the desert.

"Hey Colton," said one of the crew, "you like the show?"

Colton frowned into his empty glass. "Girls look kinda cold to me."

Someone said, "Want to go warm them up, Colton?"

Everyone laughed.

Colton stuffed a wad of chew into his lip and scratched the back of his neck.

And then the waitress came back with the food. The crew started flirting with her while she put their plates down on the table. But she said nothing until she put Colton's plate in front of him and he thanked her. "Now boys," the waitress said, "if any of you had half the manners of blue eyes over here I think you'd all get twice as lucky with the girls."

The boys hooted with laughter.

The waitress kissed Colton on the top of the head. "I could just take you home and eat you," she said.

"Whee-haw!" the boys screamed.

"Will that be all?" asked the waitress, resting her hip so close to Colton that he could smell her confusing mixture of scents, cigarettes and perfume and something like the leftover smell of a sweat-stained horse.

"What else you got?" one of the boys shouted.

"Nothin' for *you*," said the waitress, but she rested her hand on Colton's shoulder for a moment longer. Then she laughed a smoky laugh and walked quickly away to another table.

"Whee-haw!" the boys screamed again.

"Holy crap," said Colton, frowning at his double cheeseburger. "I sure hope my mom never finds out about this."

24

TRAIN STOPPING

In mid-January, six months after he'd fallen in love with her, the pretty girl in Colorado broke Jake's heart.

Colton tried everything he could think of to fix it. "Well," he said at last, "I guess my happy dance is a bit wore out."

"Just a bit," said Jake.

"She weren't a very nice girl in any case. Probably wouldn't have let you go hunting. *That* sort of girl. I heard tell of 'em."

Jake sighed.

"How about I buy us some all-we-can-eat at the Hunan Garden?"

Jake shook his head.

Jake's little brother sat next to Jake absorbing Jake's misery.

"Movies?"

Jake shook his head.

"I wouldn't mind seeing a movie," said Jake's little brother, perking up.

"I don't wanna see a freakin' movie," said Jake.

"Ice fishing?" suggested Colton.

Jake said, "I'll be fine. Why don't you boys go. Leave me here, for the love of crap."

"Fine? You don't look as if you'll be fine. You look like you need to be on 24/7 suicide watch."

Colton sat down next to Jake and that's how things stayed for a couple of hours. Colton and Jake's little brother on either side of Jake on the sofa and all three of them staring ahead, as if at a television screen. But there was no television and the only thing playing was Jake's heartbreak, on replay, over and over in his head. When it was dark, Colton sent Jake's little brother out for pizza and Mountain Dew. Colton and Jake's little brother ate but Jake said he didn't feel like anything.

"I'll have your half," Colton offered.

Jake's sad thoughts got noisier.

"Sorry," said Colton, pushing a slice of pizza at Jake. "Here."

Jake ate like he was being forced to chew cardboard. Then there was another hour of watching Jake's misery grow blacker. Colton started to jiggle his knees up and down. He cracked his knuckles. He bit his fingernails. He balled a wad of chew into his lower lip and spat. "Holy crap," he said at last, "watching your broken heart mend is like watching freakin' paint dry."

"You don't have to be here," said Jake.

"I *do* got to be here," said Colton. "In case you die of boring yourself to death."

Jake sighed more deeply than before.

Colton said, "You know what? You need to go shooting. What's in season?"

"Girlfriends?" Jake's little brother suggested.

"He-he-he," said Colton.

"Not funny," said Jake.

"No," said Jake's little brother. "Sorry."

There was more silence and more of Jake's humid sighs.

Then, "Jackrabbits!" said Colton. "They're always in season. Let's go shoot some bunnies."

"Oh crap," said Jake.

"C'mon," said Colton. "It'll fix you."

• • •

So they piled into Colton's white F150 (now that he was working on the oil patch and earning like a man, he didn't have to borrow Merinda's Escort anymore, which was nice, although, truth be told, all the boys missed the hula girl on the dash more than a little bit). They drove clear out of town and up north and turned off somewhere near the Cumberland Cemetery and then they left the road altogether and flew along in two feet of fresh white snow, the cold glittering back at the boys in the beam of their headlights.

"Holy cow," said Jake. "There's a lot of snow out here. Do you have a clue where we are?"

"Sure," said Colton.

Jake's little brother said he stopped having a clue about pretty much everything from the time they turned off the last plowed road. Snow disguised the contours of the land, heaping itself into false hills, creating a sense of solidness where there was none. "We're lost as crap," he said.

"Watch your mouth," said Jake.

"No we're not," said Colton. "Here's where all the bunnies in all the southwest of Wyoming live."

Colton kept driving and once in a while Jake or his little brother took a potshot out the window at a jackrabbit zigzagging across the crusting snow but they missed everything.

"You've lost your touch," said Colton.

Then, close to midnight, Jake suggested again that maybe it was time to go back and Colton countersuggested jumping one more drift. He said you never knew what you'd find if only you kept going and anyway it was embarrassing to go home without having hit a single bunny.

"It's late, Colt," shouted Jake over the roar of the engine.

"You're not a pussy are you?" said Colton and he gave Jake that look.

"Holy crap," said Jake's little brother.

"Here we go!" said Colton.

"Are you sure we're on high ground?" shouted Jake.

But Colton had already gunned it. The pickup caught air—the engine making a high, singing noise—and then came down up to its doors in snow. The engine stalled immediately and the boys bounced up, hit the roof, landed hard, and then there was a sense of having come to a dead standstill.

"Holy cow," said Jake. "We're not on high ground."

Colton started the engine again and tried to rock the pickup out of there, forward, backward, but nothing would give. "Okay boys, time to push."

"It's freakin' freezing out there," said Jake. He looked at what they had with them in the pickup: a couple of work gloves, denim jackets, three cans of Mountain Dew, two .22s. "We're gonna freeze to death, Colt."

Colton started singing, "If I should die before I wake, feed Jake. He's been a good dog, my best friend through it all . . ."

Jake's little brother looked white with fright. "It's not funny, Colt," he said, "who's gonna find us out here?"

"Right boys," said Colton, getting out of the pickup. He dug around in the back. "Look! A shovel," he shouted. "And Cocoa's halter. Look, and a rope."

The boys set to it. They put Jake's little brother in the driver's seat. "Gun it when we push," shouted Colton. Then they tried digging, pushing, more digging and more pushing. Jake's little brother sat up in the cab shivering with his head out the window and saying, "Holy crap, we're gonna die," over and over and pressing on the accelerator when Colton shouted, "Now!" But the wheels only made smooth, slick grooves in the snow, and the pickup sank lower into dead winter.

"Man, we're here for the night," said Jake. "We'll have to sleep in the cab."

The two boys climbed back into the cab next to Jake's little brother and warmed their hands. "Holy cow," said Jake.

"We're gonna die out here," said Jake's little brother.

"No we ain't," said Colton, "we're not even close to gonna die."

Then there was a long silence, except for the boys breathing into their hands and making squeaks of pain as the blood crept back into frozen capillaries.

At last Colton said, "I know what I'm gonna do. I'm gonna find the railway line." He got out of the cab. "You two stay put."

"Stay with the vehicle, you freakin' idiot," said Jake.

But Colton was already plowing away from the pickup in the thigh-high snow. "The railway line is close by here," he shouted over his shoulder. "I know it."

"What are you going to do?" shouted Jake out the window. "Walk to Evanston?"

"No," Colton shouted back. "I'm gonna stop the next train."

"You retard! Trains don't stop for nothin'."

But Colton had already started up the false hill of a snow-bank.

"That's how you read about people dying," said Jake's little brother.

"Come back!" shouted Jake.

Colton kept slogging up and up and away until he had walked beyond the range of the headlights and there was only a faint sense of his shape against the snow.

"I wonder if I should go after him," said Jake.

"And leave me to die here alone? Mom'll kill you," said Jake's little brother.

• • •

The brothers sat in the car, turning the pickup on for heat and then off to conserve fuel. They went through half of Dolly Parton's greatest hits that way and then suddenly there was Colton coming down the ridge again, leaping through the snow, knees coming up halfway to his chin, waving his hands and shouting, "I found the railway line! I found the freakin' railway line." He jumped back into the cab and rubbed his hands together. "It's just up there. There's a train coming. Want me to show you how to stop a train?"

"Trains don't stop for nothing," said Jake, "you freakin' retard."

"Oh yeah? You watch."

Colton started to flash the brights on the pickup, flash-flash-flash in S.O.S. pattern.

A train bore through the night, steady as a heartbeat, not slowing down, not speeding up, just da-dum, da-dum, like it was the vein that ran blood into the middle of the snow-covered plains.

"I told you," said Jake.

But then came the scream of brakes. The train's slowing. Da-dum, da-dum. Da-dum. Da-dum. Da. Dum. Da. Silence.

"Holy crap," said Jake's little brother.

"I already told you," said Jake, "watch your mouth."

. . .

The engine driver said he was sorry, but he needed to call the sheriff. Stopping the train to give a ride to three boys on the Union Pacific was, he said, highly irregular, even though he was happy to do it and it had been a pleasure meeting them and in all his years he'd never thought to find three boys stranded out here in the middle of nowhere, Wyoming. Still, the sheriff would need to be notified. So by the time the train pulled into the train station in Evanston, the sheriff was waiting. The boys got off the train into the search of his flashlight under the calmer pools of the station floodlights.

"Colton Bryant," the sheriff said. "I should have known." He shook his head. "You better tell me alcohol was involved in this little incident."

"No alcohol, sir," said Jake.

"We don't drink, sir," said Colton.

"Oh please, don't tell me you stopped a Union Pacific train in the middle of the night *sober*."

"Yes, sir."

"Then what were you boys doing? And you'd better have a good story for me."

Colton was grinning and shifting from one foot to another. "Yes, sir."

"You could have died out there."

"No, sir."

"Yes, you could."

"Yes, sir."

"So? What were you doing?"

"Looking for jackrabbits, sir," said Colton.

Jake looked up and his mouth fell open. Then he quickly cleared his throat and looked back at his shoes.

"Are you kidding me?" said the sheriff. "Jackrabbits?"

"Yes, sir."

"That's about the redneckest thing I ever heard," said the sheriff. "And that's up against some pretty stiff competition."

"Thank you, sir."

"Colton, that's not a compliment."

"No, sir."

"Oh what the heck. You boys just get outta here quick, before I change my mind and think of some of the nine hundred and ninety-nine laws you just broke."

"Yes, sir, thank you, sir."

• • •

"And that," said Colton afterward when they were back at Jake's apartment, "is how to stop a train."

Jake said, "Man we coulda died out there."

"Not as much as we coulda died back here of boredom watching your broken heart mend 'cos of whatser-freakin'-name."

"Yeah," said Jake. "What *was* her freakin' name?"

"He-he-he," said Colton.

25

COLTON AND CHASE

Winter

———— ❖ ————

Colton is a native son, so the weather and mountains, horses and guns, pickup trucks and oil rigs are what he must use to measure himself against manhood. And year by year he's growing up by this time-tested, rough-hewn method because there's truly no easy rite of passage in Wyoming. It's all bucking broncos and four-wheelers in the middle of nowhere and subzero and sheer ice and too fast everything and high, voracious winds. Sure, if you're lucky or have choices and time, there are more careful ways to measure yourself against the land than this flat-out, balls-to-the-big-sky method, but Colton doesn't see the benefit in pacing himself.

Merinda says, "I wish you'd be more careful."

Colton says, "Of what?"

"So you don't get hurt."

"Only the good die young, Merinda. That's me."

Merinda says, "Don't be such a retard, Colt. Don't say that."

Colton wraps Merinda's neck in his arms and kisses the top of her head. "Don't forget how much I love you, when I'm gone. Just remember that. You'll have your very own tough angel."

"Oh Colton, you're being a fool. You ain't gettin' out of it that easy."

"I'll be dead before I'm twenty-five," said Colton.

"You freak Mom and Tabby out when you keep saying that."

"I wouldn't keep saying it if it didn't keep being true."

. . .

And you wouldn't believe the cemeteries in Wyoming. How quickly snatched life is out here, like the sky was always too big for the earth in these high, square borders and so it *inhales* the breath of the living. Like the sky stopped being able to tell the difference between the wind on a gentle day and a person's exhalation. Take Colton and the gang of kids with whom he fished and shot geese and with whom he went hunting for nothing and everything. Out of the five of them to begin with—Jake, Cody, JR, Colton and Chase—any way you look at it, all five of them could have died twenty times over before they were out of their teens by the way they lived.

They were just ordinary, rough-broke Wyoming boys, friends through school, and then, instead of going their different ways, they all ended up on the oil patch. Even Cody, once he recovered from the realization that he wasn't going to make a pro bull rider, started working in his dad's trucking company moving water for the oil people. Then the boys—for most of them were still boys—phoned each other on their days or weeks back in Evanston and poached free nights on each other's sofas, hoping to reconnect, not so much with each other, as with the careless, innocent camaraderie of the easy years gone by.

Jake was mending his broken heart by then with a girl—a single mother of a toddler and a baby. He and Colton had known Tonya long before she got into the kind of small-town trouble that is often the burden of a pretty Wyoming girl, but it had only just occurred to Jake to fall in love with her. Their first date he said to Tonya, "Do you mind if I bring Colton?"

"Colton Bryant?"

"Yeah. I was supposed to hang out with him today."

"No," said Tonya. "That's fine."

So the three of them went to the movies, Tonya between the two boys, and then out to dinner and Tonya got into a heated discussion with Colton about the relative merits of Chevrolets (Tonya's truck of choice) versus Fords (Colton's truck of choice) while Jake built a miniature house using the salt and pepper shakers, ketchup packets, and a handful of toothpicks. That is how Jake knew he'd marry Tonya—not because she preferred a Chevy over a Ford, but because she could argue transmissions and axles with Colton until Jake said, only half joking, "Should I just leave you two here and come back in the morning when you've settled it?"

But then Jake and Tonya fell seriously in love. Now when you called Jake's cell phone it played, "Lord have mercy, baby's got her blue jeans on," while you waited for his voice to announce, "You've reached Jake. If you leave your name and number I'll git back to ya." And he spent a great deal of time with Tonya's two little children and that put Jake out of the foolin' around with his friends stakes so Colton came home on his days off and he ate some of Kaylee's meatloaf and he checked the snow to make doubly sure it was way too deep to take Cocoa out for a ride and then he wore out the bowling alley and after that, when it was good and dark, he grooved a dent in the sofa playing video games.

One evening in February, Chase phoned and said, "Colton? Don't you wanna come out and play?"

"Play what?"

"With the big boys."

"Not particularly," said Colton.

"There's a party tonight on the other side of town."

"Uh-huh." Colton tucked the phone under his chin and went back to his video game.

"Tell me you wanna drive me there."

"I don't," said Colton.

"There'll be girls for you to look at."

"I done my fill of them lately."

"There'll be beer."

"I don't drink," said Colton.

"That's why you're the perfect driver."

"Nope."

"I'll play with you at recess and give you half my lunch."

Colton said, "Nope."

"Pretty women," said Chase. "C'mon, Colton, give me a lift."

"I gotta go," said Colton.

"Okay, well, don't say I didn't try to show you a good time."

"I won't," said Colton.

"Jeez, Colton, you're gonna be sorry when I tell you what a good time I had."

Colton hung up.

· · ·

The next morning Colton woke up with a panic in his chest, like there was something or someone really important that he should have remembered. He jumped out of bed and pulled a T-shirt over his head. He went into the kitchen and phoned Tabby, "Tabby, is there something I am supposed to know about today?"

"Colton?"

"I got a bad feeling I forgot something."

"Nothing I know of."

It wasn't until lunchtime that the news got back to him. Chase had froze down walking back to his house from the party, pretty drunk in all likelihood, and he slipped and fell maybe, or stopped to rest, and the cold wind stealing in under his skin and stalling his blood and fingering its deadly way to the boy's heart so that he was solid by the time the cops got to him in the morning. It was too late for anyone to do anything by then.

And so there was Colton, driving through the icy streets of Evanston like every demon in hell was after his soul with tears pouring down his cheeks until he got to Tabby's work and he was out of the truck before it was even at a proper standstill and in front of her with his face such a pathetic mess that Tabby felt the blood leave her skin.

"Colt? What happened?"

"I should have been there," Colton sobbed. "I messed up. Oh holy crap, Tabby, I messed up. It was what I was supposed to do, and I let him down."

And nothing anyone could say to him ever persuaded Colton otherwise.

26

KAYLEE'S PHILOSOPHY
OF DRUGS

Like the way architecture has evolved differently in different parts of the world to shelter the body from a particular climate, maybe religion is the same thing for the human soul. Maybe religion is something we have constructed—like a spiritual umbrella—to shelter ourselves from the elements in which we find ourselves. So the damp politeness of Episcopalians is best suited to drizzling English summers, the Dutch Reformed Church with its bull-dog chin stubbornly facing down the southern tip of Africa, Hinduism all spice and fire in the East. Maybe, in this way, Mormonism *evolved* out in the western United States (even though started on the East Coast) to suit a place where it is all too easy for relative newcomers to let go of the earth and blow away with the wind, a land where it makes sense to set food stores aside in case of a likely emergency, a sky where a person can feel as if they have already stared down all eternity before they even get through breakfast.

Being from the West, Kaylee and Bill were raised Mormon, and although they are very private about the extent of their faith, some of the basic tenets of the church seem evident in their every-day life. An ethos of self-reliance, for example, and a strong belief in the importance of family. Also, a godly use of language and an unholy horror of intoxication. To this end, Kaylee made it very clear, if the kids ever did drugs or got themselves all drunked up

THE LEGEND OF COLTON H. BRYANT

beyond the hand of angels, she'd kick them in the pants so hard they'd be orbiting the earth for a good long while, begging not to come back down. And if Bill were to ever sit down and pour himself a liberal dose of whiskey, Kaylee wouldn't be likely to be so tolerant of that either.

But even a good rule is often better if it's broken or bent occasionally. So once a year, Bill has not whiskey, but three beers—no more or less—at the end of the hunting season. He drinks them all in one shot the night he comes home with an elk. "Where's my beer?" he asks the kids, feigning tough-guy booziness.

And Merinda and Tabby—thrilled with the novelty of it—run to the fridge and come back with three cans of Bud Light.

"One for the elk," says Bill, swallowing it down. "One for my horse," he says, drinking another. "And one for me," he says, throwing down the third. Then he takes a running start and skids across the floor on his heels, raking the wood with his spurs, straight into the kitchen where Kaylee is making supper. He pulls her away from the stove and leans her back over his arm like Fred Astaire and Ginger Rogers and kisses her full on the lips. "How'd you like your groceries delivered, ma'am? With or without the antlers?"

"William Justus Bryant!" Kaylee says.

"He-he-he," says Bill, which the kids take as a sure sign he must be drunker than a piebald judge. Then he takes a shower and goes to bed. And the next morning they eye him over their cornflakes for signs of moral deterioration. Disappointingly he is, as always, rock steady in his cowboy boots.

When they were younger, Kaylee told the kids a story about some distant cousin, or maybe (as Tabby suspects) no relation at all, but someone Kaylee had seen on television, who had dropped acid and then jumped off the roof of a multistory building to her very messy death. This death was described and redescribed in some detail by Kaylee. "Just a mess of blood and guts on the side-walk all because she was made crazy by the drugs," she says. "And can you imagine how her family felt after that? The sadness and the shame of it would be enough to kill 'em too."

And Kaylee fixes each of her children with her steady blue eyes until they nod solemnly, the fear of God branded onto their souls forever.

27

FIREWORKS

Evanston, Wyoming

It's true that everyone has a defining tragedy for their lives, but some people are unlucky enough to have a defining tragedy for every *year* of their lives and one tragedy bumps another out of the way to make way for more and more tragedies until the oases of calm between the hurt are what start to be definitive, everything else an inoculation against luck. For Melissa, it had started with her father, and the way he didn't have the sense to know where he ended and where she began in any of the important, obvious ways. So when Melissa was four, Melissa's mother took her and kept three months between herself and whatever other plans Melissa's father might have for his daughter, doing the best she could with a heart-breaking situation.

In the end, Melissa's father took the old Wyoming way out with a coward's twist—Wild Turkey and a bottle of Valium—all undone by the wind and the endlessness of it all, and maybe sick about what he'd done to his daughter or maybe just sick. But still Melissa's mother kept moving out of restless habit and working three or four underpaid jobs at once, as a registered nurse and whatever else she could find, always trying to stay one step ahead of nothing at all. It isn't just plain poverty—an ordinary lack of money—that keeps you on the wrong side of despair. It's a whole raft of poverties—a poverty of choice and a poverty of support and a poverty that comes with the certain knowledge that no one's going to take you seriously when you're invisibly decked out in an apron, working the night-shift.

Here's how patterns repeat, like your mind has wandered and you're left staring at the wallpaper in a roadside motel room, shapes going on forever until it's hard to imagine that all you need to do is get up off the crazy motif of the bedspread, open the door, and keep walking away from the sameness: Melissa pregnant at nineteen and the boy that got her into that condition halfway out the door, barely zipped up, and denying he had anything to do with it and Melissa facing part-time jobs and not much of a way to keep food in the fridge and hopelessness that looks like circles but feels more like a noose. And then one day in late June 2003 the pattern changed so suddenly that Melissa didn't realize what had happened until it was too late.

That day, Colton walked into a friend's living room and found Melissa. He took one look at her and was undone by the way she was beautiful and dark-haired and tiny and brokenhearted. So then there was a moment the size of all the sky, Colton staring at her and feeling as if someone had released an entire season's worth of geese below his ribs and Melissa thinking, "Well?"

And she may have even said it because Colton said, "What?"

"I'm Melissa."

"Colton," said Colton, snatching his ball cap off his head. "I was . . . looking for Ruth . . ."

"She's gone to get some hot dogs," said Melissa.

"Oh," said Colton.

Melissa lit a cigarette and smiled at him a little, most of the smile going into her eyes. Nathanial got up from where he had been lying in front of the television and wrapped himself around her leg.

"Hi there, half-pint," said Colton.

Nathanial pushed his stomach out at Colton.

"He yours?" Colton asked Melissa.

Melissa nodded. "It's just him and me," she said.

"Ah," said Colton.

Nathanial started to spin around with his arms out on either side. "I'm busy, I'm busy."

"You're *dizzy*," said Melissa, "silly cowboy."

Colton folded himself down. "Want to know something, little dude?"

Nathanial stopped spinning and frowned.

"I had a real wild horse one time," Colton said, "and I called her Cocoa and I trained her all by myself. But she ran away."

Nathanial crossed his arms over his chest and summoned up a look of deep concentration.

"I looked all over for her. Then one day she fetched up in Freedom, Wyoming."

"You've got to be kidding," said Melissa.

"I ain't kidding." Colton stood up. "Fetched up at an auction. Some guy trying to sell her as his own."

"Horsey!" said Nathanial.

"Brand inspector called my mom and said he'd found her registered to us. Saw her BLM brand."

"Horsey!" said Nathanial.

"She was gone a whole year," Colton said. "I nearly killed myself and my best friend trying to find her and all along she's in a domestic herd out there." Colton waved north. "Fat as butter by the time Mom brought her home."

Nathanial lost interest and went back to the television.

Melissa took a drag off her cigarette and eyed Colton through the smoke. She said, "She should be back any minute."

Colton frowned.

"Ruth," said Melissa. "She just went down to the Loaf 'N Jug."

"Oh," said Colton. "Oh."

"Yeah." Melissa smiled and smoked.

There was a long pause.

"She's a great horse," said Colton then.

"Oh?"

"I mean, I'd put a kid on her."

"That's nice."

Colton looked at his hands, his nails bitten down to flat nibs. "I don't think I've seen you before," he said.

"I've moved around." Melissa put her cigarette out.

"Not me," said Colton. "Aside from I've been three years in the Upper Green flow testing."

"Flow testing?"

"Yeah, but I'm gonna drill again someday soon. You know, out on the rigs."

Melissa smiled. "Yeah," she said, "I know what out on the rigs is."

"Yep," said Colton. He stared at Melissa for a few moments more. Then he cleared his throat. "Look, do you want to come and shoot fireworks with me?" he said.

"What?"

"And Jake."

"Who?"

"We've got artillery shells and cannons . . . We went to Porter's and bought half the store. It'll be better than the fireworks show in Vegas. You'll see . . ."

"Vegas?"

"Do they have fireworks in Vegas? That's if they have fireworks in Vegas. Yeah they do, every night, don't they?"

"I imagine," she said.

· · ·

So for Colton and Melissa's first date, Colton and Jake stood at the bottom of Jake's garden and let loose with four hundred dollars' worth of fireworks the sound and size of bombs like they planned to win the war. Colton was showing off for Melissa, of course, holding on to the cannons until the last possible moment and almost setting himself off with them. And then it was Jake's turn and Colton said, "Come on, you pussy, hold on to it," and by the time Jake let go, it was too late and the shell went off at an odd trajectory, shot off the neighbor's roof, skidded over the top of their garage, and smacked into a cop car. Colton rocked back on his heels and laughed and his head was thrown back and his hands dangling by his sides, "He-he-he!"

"Holy crap, it's a lady cop," said Jake. He looked at Colton. "They're the worst."

"He-he-he," laughed Colton. "What kinda bird doesn't fly?"

Melissa put out her cigarette. "Oh no," she said.

"A jailbird," said Colton. "He-he-he."

And then there she was, serious as a heart attack and arms cocked like cormorant wings over a gun and cuffs, the lady cop. "Did you *intend* to shoot me?"

Colton said, "Evening, ma'am. No, ma'am."

For a good ten minutes the lady cop read them the book, on and on she went, and Colton just hit her with those cornflower blue eyes of his and said, "Yes, ma'am," and "No, ma'am," and "Most definitely sorry, ma'am," and finally you could see the lady cop get soft and her shoulders relaxed. She said, "You never let this happen again."

"Never, ma'am," said Colton, and he shot her a smile to light up the sky.

She shook her head. "You should be careful what you do with those eyes," she said. "Someone might get hurt." And then she was gone, back to her car.

"Holy crap," said Jake. "Did she just say that?"

"Say what?" said Colton.

Melissa looked at Colton like she was seeing him for the first time. "Holy cow," she said.

28

DRIVING ALL DAY

Wyoming/Utah/Arizona

———◆———

Melissa herself can't put her finger on why she woke up one day, a month after the night of the fireworks, and decided to pack up Nathanial and everything she owned, which was barely enough to fill the trunk of her car, and drive out of Evanston, more or less straight south until the landscape looked entirely different, all empty and red and low, like all the earth was being swept toward the Grand Canyon and no Wyoming chill in the air, the kind you get as the sun sets. She didn't tell Colton, or any of her friends, that she was going, because she didn't have time and she didn't know herself that she was leaving until she was already gone.

Nathanial was still in his pajamas when she buckled him into his car seat. He started crying. "Cowboy up, cupcake," she told him.

"Why?" he said. "Where we going?"

"I don't know," said Melissa, lighting a cigarette and pulling out into the road. "I don't honest-to-God know."

It didn't make it any easier when, halfway to Camp Verde, Arizona, Melissa recognized that what she was running away from was a dead man. The living you can press into corners, see with your eyes, touch with your fingers, so if you're cunning and quick, you can always run away from the living, but the dead are always with us and from them there is no escape. Then she was crying and Nathanial started up again and Melissa was saying over and over, "I'm sorry, cupcake. I'm so sorry."

But it was too late to turn back by then. They were halfway to somewhere else and they'd already burned the gas it took to get there. And then, once they got there, there was nothing much to keep them in Camp Verde, Arizona, and there was everything to turn around for—the beginning of free-falling in love with Colton and the support of her friends and the familiarity of Evanston—but Melissa had pulled into town with a drop and a half left in the tank and it was dark and her choices weren't looking so multitudinous. She spent her last thirty-five dollars on a room for the night and the next morning she found a place to put Nathanial during the day and she found a minimum-wage job as a clerk in a gas station.

So it went for a couple of months. By the time she had paid for babysitting and rent, and gas and diapers and baby food, there was nothing left for her to eat. So she lived off Mountain Dew and cigarettes, the combination of which, in sufficient quantities, can stave off everything from despair to hunger in fifteen-minute increments. And every night she curled up with Nathanial in his dinosaur pajamas and she sang, "But Momma kept the Bible read and Daddy kept our family fed." And when Nathanial was breathing steady, his even breath against her neck, Melissa cried herself to sleep.

29

PATTERSON-UTI DRILLING

Upper Green River Valley

By about now, Jake was getting ready to move on from flow test-ing. He'd been offered a job making decent pay selling service supplies to wells. And with Jake gone, and with three years between the fall off the rigs that broke his foot and a fresh start, there was really nothing else keeping Colton at the flare pits because he'd been ready to move back onto the rigs from the day he left, so he reapplied for a drilling job with a company other than the one that had laid him off.

It had turned cold the day Colton drove up to Casper in the heart of Wyoming's oil and gas country, fall laid a frigid breath over the filtered sun. It was the hunting season of Colton's twenty-third year and the scented wind off the mountains and the bittersweet smell of turning aspen leaves made him itch to be up at the snowline tracking elk into deadfall. He drove with one arm out of the window, wound down, the better so he could get a faceful of Wyoming air. He had the old Neil Diamond CD on the soundtrack and the last of the summer's geese were veeing south. Colton stuffed a wad of chew in his lip and hummed along.

The drilling companies had taken over a hotel in the middle of Casper. Colton filed into the lobby where tables were set up like the army signing up boys to be soldiers. He joined all the other potential recruits filling in a questionnaire of yes-or-no questions. Are you able to work away from home for extended periods? Are you able to withstand an all-weather, outdoor job that is very demanding physically? Are you able to work a significant amount of overtime? Are you able to work weekends and holidays? Colton colored in all the circles next to the line that said yes.

"Heck, yes," he said.

The piss test was the next thing he had to do. He showed the tester the skin of his stomach—lifted his shirt, opened the waistband of his pants—to prove that he wasn't carrying clean urine into the cubicle taped to his body in a condom or in some kind of prosthetic device. (The Whizzinator comes with synthetic, dehydrated urine and is available in white, tan, Hispanic, brown, and black. According to the manufacturers and the testimony of scores of satisfied customers, it has the look and feel of a real penis.) The cubicle didn't have a faucet or a sink. There was no water in the lavatory. Nor was there much in the way of privacy. Colton pissed in a sterile catcher and handed it to a woman in a lab coat.

"Name?"

"Colton H. Bryant."

"Date of birth?"

"June 10, 1980."

"Okay."

"Do I have a job?" he asked.

"That's not for me to say," said the woman. She was bored, overweight, and overdue for a cigarette break. She checked the label on Colton's jar and made a mark on the piece of paper taped to the clipboard.

Colton said, "You wouldn't think it would be so difficult to be oil-field trash."

The woman looked up and found herself on the receiving end of the most startling blue eyes she had ever seen in her life. "You like hard, dirty work?"

"Love it."

The woman was smiling now. "Boiling hot or freezing cold and a wind, always blowing sideways, day and night?" she asked.

"Love it," said Colton again.

"Oh yeah?"

"Heck, yeah."

The woman lowered her clipboard and cocked her hip at him.

"Third generation on the rigs," said Colton.

"You're a Wyoming boy then?"

"Yes, ma'am," said Colton. "That blowing sideways wind, she's my theme song."

The woman laughed. "One of our crazy-assed cowboys," she said.

"That's right, ma'am."

"You boys sure keep this damn country running."

"Thank you, ma'am."

"No," said the woman, "thank *you*."

"He-he-he," said Colton.

"Oh well," she said. "If your test comes back clean and you're half alive in every other way, then yeah, they're not so picky. Sure, I would say you've got yourself a job."

Colton put his arm over the woman's plump shoulders. "Whee-haw."

"All I'll say," said the woman frowning down at her clipboard, "is get yourself home alive at the end of each hitch." She cleared her throat and looked up. "I bet you have someone waiting for you, don't you?"

"Not yet," said Colton. "But I will. Sure I will."

The woman watched the young man leave the building. A tall boy with wide shoulders and funny half-dancing gait, like he favored his left foot perhaps, and as if his arms and legs were attached to the sky by strings. Then automatic doors opened to the grey world outside and the boy stepped out in the parking lot and the wind picked up a ground blizzard and swallowed him up in it. And then there was another recruit in front of her and a fresh jar of urine to be processed. She sighed.

"Name?" she asked.

. . .

A day later, on October 17, 2003, Colton was hired as a floor hand for Patterson-UTI, drilling for natural gas in the Upper Green River Valley. He was given the company's New Short Service Employee Orientation, the quicker to get boys out of the classroom and onto the rig. But even obsessive training—tongs-use training, scrubbing training, swinging-pipe training, blood-borne pathogens, hazard communication, fall protection, confined-spaces instruction, respirators training, lockout/tag out, forklift training—won't stop an accident, because although everyone talks about safety, no one means it as much as they mean money. Whatever they say, what they really mean is, "Safety, safety, safety, have a slice of pizza. Safety, safety, safety, have a doughnut. Safety, safety, safety, now go out there and get as much gas out of the ground as quickly as you can and make us all rich as shit without killing yourself."

It's a good bet, Colton was halfway out of his chair and halfway onto the rig before he even heard the last bit.

30

DRIVING ALL DAY
AND NIGHT

Wyoming/Utah/Arizona

————

They say Wyoming is like a small town with a really long main street. That is why, for all its appearance of emptiness and live-and-let-live ethos, the state has a small town's propensity for taking care of its own. You have only to see the notice in a post office announcing an illness or death to see what it is to live in a small-town state—the casseroles and meatloaves and cakes and the people bringing in your hay, feeding your horses. It's enough to restore your belief in humanity.

The flip side of this small-town state is its habit of breeding vicious gossip and for paranoia and for neighbors shooting neighbors or family unloading into family over something that happened so long ago that no one can really remember all the details now. And love-hate, bittersweet opinionated is as easy as taking ownership of the wind—everybody does it, without even realizing it. And everyone knows everyone else's business, and what they don't know about someone else's business, they'll fabricate. So it didn't take long for it to get back to Colton, via Ruth, via goodness only knows who, that Melissa was living off Mountain Dew and cigarettes in Camp Verde, Arizona, and that she was unhappy and that she wished she'd never left Wyoming. And also that she'd been asking about Colton.

Which was all Colton needed to know, true or untrue.

He didn't even sleep on it. He left Evanston at two in the afternoon in his white F150—"My knight in a white Ford," Melissa called him afterward—and he drove six hours, clear through Utah, and when he got to Page, Arizona, it was eight at night. By then, Colton was all hopped up on chew and Mountain Dew and he needed gas. So he filled up and then found somewhere that would serve him a sit-down meal. He ordered a double burger and fries from a waitress at a joint where neither the plants nor the utensils were plastic and he phoned Melissa from a phone booth outside. Fall was coming fast, but it was still warm here, the asphalt smelling of lost summer and the air burnt with cigarette smoke and oil and old, shredded rubber.

"M'issa?"

"Yes."

"I'm on my way to Camp Verde," said Colton, forgetting everything that he had been rehearsing for the last six hours. "To fetch you home."

"What?"

Colton looked at the sun and made a quick calculation in his head. "I'll be there by midnight, I reckon," he said. "Another four hours should do her."

"Colton, what are you talking about?"

"M'issa, how about you marry me?"

There was a very long pause.

"I'll take Nate as my own son." His own voice sounded long and lonely going down the wire. "M'issa, you there?" Colton tapped the receiver. "M'issa?"

"Holy cow, Colton Bryant," said Melissa, "are you out of your ever-lovin' mind?"

"No, I ain't," said Colton. "You can't live off Mountain Dew and cigarettes for the rest of your life."

There was another long silence.

"I got me a job drilling in the Upper Green," said Colton.

"I'm back on the rigs. It's real money M'issa, sixty grand a year if I work overtime. But you got to come back with me now. I've got another hitch coming up in a couple of days."

More silence.

"I'll take care of you, you'll see."

Melissa still didn't say anything.

"I'm not much of a one for fighting. But I'd knock a man on his ass if you needed me to. I really would."

Still nothing.

"I'll teach Nate to ride and hunt. It's never too early to get a boy on a horse. He can use Cocoa. She's quiet as an angel now."

The sun beat long rays across the tarmac into Colton's eyes. He turned his back to it and waited until he couldn't stand it for another moment, then he said, "I'll take that as a yes, then."

Melissa shut her eyes against the sensation that her chest would burst. "Yes," she said. "Yes, Colton H. Bryant, you can take that as a yes."

Colton hung up. "Whee-haw," he said softly.

. . .

At that moment, the waitress of the burger joint looked out of the window at the parking lot and saw her customer with the bright blue eyes in his cowboy boots and Levi's and a black ball cap with the words western petroleum written in orange letters across it doing some kind of crazy dance all around the phone booth, like his arms and legs were attached to the sky by strings. When Colton came back inside to get his meal, she said, "You win the lottery out there or something?"

Colton was squirting the better part of a bottle of ketchup onto his plate. "Or something," he said, grinning. "Yes, ma'am, I most certainly did, ma'am. I just did a whole lot better than won the lottery."

31

MARRIED

Evanston, Wyoming

They were married a month later in the courthouse in Evanston on November 10, 2003, a dirty, cold day, piles of grey snow banked up on the winter-dreary lawns around town. Someone in the courtroom—a receptionist perhaps, or the security guard—took photographs that show Melissa, tiny and dark-haired and smiling next to Colton, who is wearing a pressed white shirt, a clean pair of beige jeans, one of Bill's saddle-bronc belt buckles, and cowboy boots. He's slicked his hair down for the event, but tufts of it have escaped the comb and are poking up in rebellious spikes. He is grinning goofily, which makes him look even younger than his twenty-three years and five months. After the wedding, Colton and Melissa went to the Sinclair station where Tonya was working as a clerk and Jake was using his week off from the wells to hang out with her.

"Hi kids," said Colton, still with the goofy grin pasted all over his face.

Jake looked up to see Colton citified up and spit and polished. "Colt, what they do to you?"

Colton grinned wider and squeezed Melissa's hand. "Guess what we just went and did?"

Jake and Tony looked at each other.

"Well?" said Colton, hopping from one foot to the other.

"You just went and got married," said Jake.

"Hey, how'd you know?"

"Lucky guess, I guess," said Jake.

"We wanted you to be the first to know," said Colton.

"Why didn't you invite us?" said Jake.

"We didn't invite anyone."

"No one?"

"Nope."

"Holy cow," said Jake, "you didn't invite your mom and dad?"

"Nope."

"Have you at least told 'em?"

"Nope," said Colton.

"Holy cow, Colt, you'd better go and tell her right now. And Lord have mercy, she's gonna want to tan some hide." And then Jake thought about Merinda and Tabby and he said, "And your sisters are gonna want a piece of your hide too."

Melissa's smile wavered a bit.

"They'll find out soon enough," said Colt, putting his arm around Melissa's shoulders. "Mind over matter, kids, mind over matter." Then he said, "Come on, let's go out and get us some Chinese food. I'm starving and I'm paying, so let's eat."

So the four of them celebrated at the Hunan Garden on Front Street just a block down from the pawn shop where Colton continued to do a pretty steady trade every month or two with his custom-made saddle or his cowboy boots or a gun. It was like he figured it was free money to hand over a couple of his belongings and walk out with cash.

"Check that out, kids," said Colton, pointing to the Uinta Pawn shop, "'Colton's Closet' is what they should name that place. Heck, it must smell like my boots in there by now."

They ordered sweet-and-sour pork and dumplings, Szechuan beef with broccoli, and chicken chow mein washed down with refillable Mountain Dew and Diet Coke until Colton sighed and pushed himself back from the table and said, "If I have another bite, I'm gonna have to be carried out of here in one of them stretchers."

"You got to eat your fortune cookie at least," said Melissa.

"Read mine for me," said Colton.

"Oh please Lord don't tell me I married myself an illiterate," said Melissa.

"Close enough," said Jake.

Melissa cracked Colton's fortune cookie then opened the piece of paper that fell out of it. She frowned and looked up.

"Go on," said Colton, grinning, "what'd it say?"

"It's blank," said Melissa. "Look." She showed Colton the plain white slip of paper.

There was a moment of silence and then Colton said, "It don't matter. That way I get to write my own fortune."

"Which is?" said Jake.

"Mind over matter," said Colton.

Jake laughed. "I coulda told you that."

"I don't mind," said Colton, "so it don't matter."

32

DRILLING

Two days after they were married, Colton drove the two hours back up to the man camp in Big Piney and by six o'clock that night he had caught a ride out across the high plains for his shift on the rig, which seemed tiny from afar, seen under the shadow of the Wind River Mountains. The high plains have a way of diminishing and distorting the scale of everything, and until you've climbed the stairs to the doghouse on a drilling rig, it's difficult to imagine the height of it—fifteen stories, all told—or the sense of exposure out here; it's as if the tower were anchored to a swell of water that might shift at any moment and set you adrift.

And it's not just that a rig is vast, but it's ingeniously brave. The courageous imagination it takes to bolt yourself onto the high plains and drill down day and night, following a map of the world no one can see, but that geologists can track and picture, plates and plates of everything this earth has been laying down, trapping natural gas in pockets like a cross section of a bowl of Pringle chips. The driller and the tool pusher coordinating the movement of six men, like some kind of rough man's dance so that no one gets behind a fast-moving pipe, or a quick pair of tongs. The drill spins into the ground at such a speed that a driller's helper in Gillette, poorly trained in a prisoners-to-work Volunteers of America program, caught his right hand in the drill, instinctively reached with his left hand to free himself, and had both arms ripped off at the sockets. He died on location.

. . .

Colton rubs his gloves together and breathes into his sleeves. He can't feel his fingers. A deep, killing cold has settled onto the high plains. The snow veers off the scraped-over sage in shiny silver sheets, turning air to metal by an enchanted catalyst of winter, but Colton and the other men move with a kind of casual ease around the rig, so many grey shadows, tiny as sailors on a battleship, brave and competent and powerfully equipped. Twelve hours at a time, counted down in measurements of feet bored hotly into the reluctant ground. The wind, cold and picked up off the plains like this, is old news for Colton. The smell of the mud they use to lubricate the drilling bit reminds him of Bill's greasers lying by the door on the front porch at home. For Colton, the oil and diesel and metal are childhood scents.

There's a safety meeting in the doghouse before any drilling could start, routine as a morning prayer at a monastery that went something like this. "Let me explain something to you real simple, which is this. The quicker we get 'er drilled the less it costs. The less it costs, the more money is to be made. Drill it quick and drill it straight and don't get blown up doin' it. Get it?"

"Got it."

"Okay, sign here, boys."

Colton scribbles his initials on the bottom of a piece of paper.

"Time is a whole lot of bank."

"Yep," says Colton.

"Then get on with it, boys."

"Yep," says Colton.

"Straight and quick like you mean it."

And Colton is gone like a shot.

33

THANKSGIVING

Evanston/Rawlins

———◆———

There aren't, but there should be, a hundred different words for wind in Wyoming; crop-burner, roof-lifter, barn-raiser, widow-maker. They say that one summer, in 1968, up near Chugwater, the wind stopped blowing for a few moments and spooked the horses. And every year somewhere between Rawlins and Laramie, the wind flicks over a couple of freight cars on the Union Pacific line. Wind, day and night, taking the names of men, women, and children—you and you and you, the wind says—and picking them off, one at a time. Off snowmobiles and oil rigs, off icy roads, it's the wind. And it's the wind blowing them over into cold water or pasting them onto the high plains in midsummer. At other times, the wind upsets forklifts or unloads guns, opens up a badger hole in front of a horse or sends down a bolt of lightning. Surely, to be born to such a wind is to be born half-given back to the earth. Untethered from the get-go.

That day, the day before Thanksgiving in 2003, the wind seemed to have Jake and Tonya's names picked out. They were on their way to Denver—Tonya had a job hot-shotting a drill bit from here to the airport en route to an oil company in Italy—when the wind came in busting a gut, tunneled just here into more than ordinary fury. It came from behind and caught the back of Jake's pickup truck and flicked it right about. The front wheels touched black ice and the pickup came off the road and tumbled like a toy. In eight separate frames, slow as it takes to think about life and dying, Jake saw a billboard that advertised thirty-five-cent cones with a picture of an ice cream and the name of a hotel in Cheyenne and he was thinking, "That's some cheap ice cream." What he said aloud was, "Holy crap. Hold on, Tonya! Holy shit, oh shit!"

And then he felt the truck hit the ground on the driver's side and his head connected with the gun rack and after that he didn't know anything else for quite some time. Tonya counted the truck rolling five times. And then she lost count of almost everything except the world spinning and spinning. And then, at last, the violent velocity rocked silent. Here was Tonya's heart stopped too, and the blood pooled in her neck like a crashing rodeo rider and she found she couldn't breathe. And then there really was silence, and the two of them hanging upside down from their seatbelts.

Jake's eyes were closed and blood was pouring from the top of his head. Tonya tried to reach over to him but her seatbelt stopped her and she couldn't unbuckle. Jake opened his eyes and looked over.

"You're alive," she said.

"Looks that way."

"Your truck," said Tonya.

"We'd better find the cell phone and get ourselves outta here," said Jake.

"Are you gonna kill me?"

"Why?"

"Your truck," said Tonya. "I rolled it."

"It don't matter. It's just a stupid truck."

"It's your baby."

"No," said Jake. "*You're* my baby. Okay? You gonna be okay. I'm gonna take care of you."

"Jake," said Tonya, starting to cry. "Holy cow, Jake."

"Where you bleeding, girl?"

"I ain't bleeding."

"Then where's all this blood coming from?"

"Jake," said Tonya. "It isn't me. It's you."

And then there was a Mexican couple and a trucker trying to wrench the door open and when they got Tonya and Jake out, the Mexican man had a Saint Christopher held up, as if in blessing, over the two of them. He was grey and his hands were shaking as he was praying, but then the woman pushed him to one side. "You're not helping," she told him, and she threw a blanket over Tonya's shoulders and held on to her and led her to the side of the road and another woman had Jake sitting down on a winter jacket and the trucker was trying to stop the bleeding on Jake's head with a T-shirt.

"You gonna be fine, man," said the trucker. "The ambulance is gonna be here." He looked up at the sky and said, "We could do without the wind, though." He was shaking too.

"Yeah, this wind," the Mexican man agreed. "It'll make you crazy."

"What you carrying in the box?" the trucker asked Tonya.

"A drill bit," said Tonya.

"It looks like a bomb," said the Mexican.

"It's a drill bit," said Tonya.

"Lucky for you it didn't go off," said the trucker.

"Be quiet, you. You aren't helping," said the Mexican woman. She tugged the blanket around Tonya's shoulders a little tighter. "You don't talk or nothing, lady," she told her. "You take it nice and easy."

. . .

This being Wyoming, by the time the story hopped its way through the nervous crackling of bad cell phone connections to Evanston it got to Colton (playing video games on Cody's sofa) that Jake and Tonya were in the Rawlins emergency room and he had been decapitated and she had broken her neck. That was all he needed to hear. Colton was off Cody's sofa and into Merinda's

Ford Escort in seconds. He burned home doing about ninety and ran into the house.

"Hey," said Melissa, "where's the fire?"

"Jake," said Colton.

"What?"

Colton's face was completely wet with tears. He snatched money off the dresser. "Jake's been hurt."

"But . . ."

"I gotta go!" shouted Colton.

"Holy cow," said Melissa. She stood up. "I'll come with . . ."

But Colton had already run out the door, tunnel vision for Rawlins. He made that drive, from Evanston to Rawlins, in an hour and fifteen minutes instead of three, and he would have been even faster but the governor on the Escort kept kicking him back every time the needle on the speedometer went past the end of where the numbers are written. So the needle quivered over to the extreme right, the governor kicked in, the needle sank back to about forty. Colton gripped the steering wheel. "Come on," he told the car, "you crappy little piece of shit, don't slow me down."

Colton left the car running, the door open, made it across the parking lot in about eight strides, and ran into the waiting room at the Rawlins hospital and there were Jake and Tonya sitting on orange plastic chairs facing the parking lot, not knowing who was going to pick them up. Jake had a big bandage on his head, like a mummy. Tonya was in a neck brace. Colton ran down the corridor, tears streaming down both his cheeks.

Jake stood up. "Colt?"

Colton slid the last few feet on the heels of his cowboy boots. "Come here, you two," he said. "I thought I'd lost you." And then he had Jake and Tonya in his arms. "You're alive, you sons-a-guns. You're alive." He looked at the bandage on top of Jake's head. "Man, you ain't been decapitated. You just been scalped."

"I'm fine," said Jake.

"Man, I heard you was dead." Colton kissed the top of Tonya's head. "But you're okay, ain't ya?" And then he kissed Jake's band-ages. "And they told me you was dead," said Colton. "Man, I thought I'd lost you."

. . .

Aside from being the site of Jake and Tonya's accident, Rawlins is home to the Wyoming State Penitentiary, historic and modern. The old facility shut down in 1980 after more than a hundred years of service. Butch Cassidy did time there for horse theft and Big Nose George was hanged from a streetlamp out front by a group of citizens who broke him out of the place to give him a taste of Wyoming justice. The small-time outlaw had made the mistake of killing and dismembering both the popular sheriff and the popular deputy sheriff of Rawlins in the late 1870s and then of attacking the jailer. The Rawlins doctor, John Osborne, who pronounced Big Nose George dead, had the corpse skinned and sent the resulting hide down to Colorado to have it tanned, out of which he had a pair of shoes, a doctor's valise, a vest, and a purse made for himself. An ashtray was made out of the top of the outlaw's skull. The doctor went on to become Wyoming's first Democratic governor and he wore the shoes made out of Big Nose George's hide to his inauguration ball. The governor's shoes lie under glass to this day at the Rawlins Museum where you can also see Big Nose George's death mask and other fantastic exhibits.

So Rawlins's economy is tied hard to the incarceration business, and imagine a prison in the long, sweeping openness of Wyoming. All that big sky and accumulating miles of sage piling space upon space, the trains running through here screaming free-dom every time they stop and go. But there's nowhere to hide, even if you could get out, the sentries can see as far as eagles from their lookouts. Rows of low, heartbroken hotels accommodate the relatives of inmates and the restaurants cater to the summer Outlaw Trail tourists and it's all as if a high wind might blow the place back into the Old West days of gunsmoke and high noon shootouts.

Of course, Colton knows the burgers anywhere within about a five-hundred-mile radius of Evanston, even in Rawlins. So he takes Jake and Tonya to a joint that sells the Outlaw (a one-pound burger, fresh Angus ground beef), the Lifer (a double cheeseburger), the Sheriff (same as the Outlaw except with bacon and jalapeños), and so on.

"My best friend here got decapitated," Colton tells the waitress.

But, being from Wyoming, and especially being from Rawlins, she's heard it all. "Is that right?" she says, bored. "And you want fries, baked potato, or coleslaw with that?"

And then they get on the road back west and they putt along going about sixty so as not to freak out the accident victims, no one saying much of anything until at last Colton says, "How are my little people doing?"

Jake says, "Head aches like a sonofa."

"Did they give you something at the hospital?"

"I wouldn't take it."

"Sheesh," says Colton, "even I'm not that much of a retard."

"I don't like pills."

"Me neither, but, there's a time or two a body just needs a little help."

"You're one to talk."

"We'll stop and get you some Tylenol next gas station."

So at the next gas station Colton and Tonya left Jake in the car and Colton led her conspiratorially to the medicine aisle and searched around for a bit until he found what he was looking for. "Let's slip him one of these things," he said, holding up a bottle of Tylenol PM.

"Tylenol PM?"

"Man, these things will knock your lights out. But we'll pretend it's ordinary Tylenol."

"Okay," says Tonya.

So they gave Jake a Tylenol PM. "Some people eat these by the handful," says Colton. "So you can take two."

"One'll do her," says Jake. He takes it and Colton winks at Tonya. They bump back onto the highway and before Wamsutter, Jake is passed out and he sleeps the whole way back to Evanston.

"What did I tell you?" Colton says to Tonya. "Knock you on your ass, that Tylenol PM stuff."

34

A Serious Life

Colton called and said, "You wanna go out shootin' bunnies with me?"

"Are you kidding?" said Merinda.

"Nope."

It was January and the town was blustered half to death with blown-out, raggedy Christmas decorations and everything was glutted and hungover with the holiday season and still the bigger part of winter had to be faced. Nights were coming early and dirty, staining streetlights, casting no shadows into the disappointment of itself. Days hung low and short, like they could barely stand to be here at all and there was nothing for a body to do except wait until the days grew longer and the sky lifted and the sun came back from the south.

"Bunnies?" said Merinda.

"Yep," said Colton.

Merinda sighed. "What is it, Colt? It's blowing sideways out there."

"That's because this ain't Florida," said Colton.

"Can't you tell me on the phone?"

"Nope."

"Okay," said Merinda. "I'll meet you at the Maverick."

"I'm already there," said Colton.

Sure enough, Colton was standing up against his white F150 in the parking lot of the Maverick, the snow blowing off his Carhartt jacket, chin down against the wind, cowboy boots heel-dug into the ice-covered asphalt. Merinda pulled up in her little blue Escort with the hula girl on the dash. She switched off the engine and got into the passenger seat of the pickup truck. Colton climbed in behind the wheel and they were out of the parking lot before Merinda had her feet all the way pulled up into the cab. "Don't worry if you run over my feet," said Merinda.

"I won't," said Colton.

And that was all that was said all the way out to the Cumberland Flats. Then Colton switched off the truck, reached back for his .22, and put the gun across his lap.

"There aren't any bunnies out here," said Merinda.

"Nope," said Colton.

Merinda looked out the window at the way the snow-covered ground reflected brighter than the sky, turning the world into a silver basin, the odd star poking feeble winter light down from the filthy sky. And that's how the world stayed for some time—no bunnies and all the silence a person could stand—a careless, unpeopled silence that went back to before there were words for it.

Then, "She's pregnant," said Colton.

Merinda said. "I know."

"How'd you know?"

"I just knew."

Colton nodded.

"That's great news," said Merinda.

Colton's hands clenched over his gun, "Yeah. I wanted you to be the first to know."

"Thanks."

"It's all kinda quick," said Colton.

"No kiddin'."

Colton leaned his head on the steering wheel. "I've never been happier," he said. "It's just . . . You think the kid's gonna be okay?"

"Of course it's gonna be okay."

And then suddenly Colton burst into tears. "What if he's like me? What if he gets teased and he has to struggle like I did?"

"Colton, what are you talking about?"

"What if he's a slow learner? Merinda? What about that?"

Merinda smiled. "What if he's a she?"

"What?"

"Colt," said Merinda, "you're the kindest, most forgiving human being I know. You're the best uncle on the planet to Preston and Tabby's kids. You love your nieces and nephews. You're gonna have a beautiful baby. You're gonna love it."

Colton sniffed and wiped his nose on the back of his jacket.

"Unless you teach it to wipe its nose on its sleeve."

Colton laughed. Then he said, "It's all the Mountain Dew, I guess."

"What's all the Mountain Dew?"

"Babies is."

"What are you talking about?"

"Keeps me awake all night."

"Oh brother," said Merinda.

"He-he-he," said Colton.

Then neither of them said anything for a long time until Merinda said, "You're gonna be great with a baby."

Colton pressed his knuckles into his eyes. "It's all I ever wanted." He looked at his sister. "Messin' around days is over, Merinda. It's gonna get more serious from here on out."

Merinda nodded.

"Another kid. My own baby? Holy cow." Colton took a deep breath, then he reached down and started the pickup truck. "You think I hunted all the bunnies there are out here to hunt?"

Merinda said, "Pretty much."

"That's what I thought."

Colton eased the truck back into a clearing where he could turn around and get back onto the highway. "Just so you know,"

he said, "if something happens to me and M'issa, we want you to
have the kids, okay?"

"Okay," said Merinda, "but nothing's gonna happen, Colt. It's
gonna be just fine."

"Sure," said Colton, "It's gonna be just fine."

35

MARRIAGE AND
ROUGHNECKING

Evanston, Wyoming

———◆———

Roughnecks live in half-month or weekly increments, hoping personal time pans out around their one or two weeks away from the oil patch. There's no such thing as sick leave, paid leave, paternity leave, compassion leave. Either you're there for your hitch or you're not, no one's going to keep your place warm for you while you coach your wife through labor or your kid through a baseball game. There is a mechanical dispassion to the process of being hired and fired. The rigs are the only constant. They have to keep drilling hour after hour—storm, heat, sleet, ice, sun—no matter what. They'll slap another beating heart on the rig to take your place if you're so much as five minutes late.

So when Jake and Tonya were married the summer after their car crash, Colton stayed up thirty-six hours straight to make it as best man to the wedding and still work the two twelve-hour shifts on the drilling rig on either side of the wedding. He left Big Piney as soon as the crew got back into the man camp at seven and he was in Evanston by ten, cutting it fine, because for one thing, it took him the better part of the rest of the morning and an hour of the afternoon to get scrubbed up out of his greasers and into a tuxedo rented from the Bridal Center in Salt Lake City—black bow tie, black waistcoat, black coat, and a blue corsage. And then the church by three, and the reception after that, with all the speeches and cake-cutting.

The music went on until midnight, Colton slow-dancing with Melissa, who was three months shy of delivering Dakota. When the bridal couple left the reception for their honeymoon, Colton hugged Jake, both of them awkward and a bit rigid in those silly outfits, and he kissed Tonya stiffly on the cheek. After that, he drove Melissa home and she took a bath while Colton checked on Nate. Then he rubbed Melissa's back, and once she was asleep, he got on the road north and hurried back to be at the man camp by four-thirty to catch the shuttle to the rig at five with the rest of his crew.

A week after the wedding, Jake gave Colton an official portrait of the two of them, framed. It shows a background of trailing white cloth over fake Ionic pillars and the two men in the foreground, groom and best man, smiling a little uncertainly at all the unaccustomed pomp—the stiff monkey suits, the flower arrangement poking out of a hole in their suits, the slicked-down hair—but proud too, as if they had achieved some kind of countable milestone. And underneath this portrait, on the white border left by the photographer, Jake has written in an uneven, blocky hand, "best friends forever."

36

THE DEATH OF
LEROY FRIED

Upper Green River Valley

On Monday, August 2, 2004—a sunny summer afternoon, not too hot and the wind becalmed to a gentle breeze for once—a forty-nine-year-old roughneck was killed by a falling support beam in the Upper Green River Valley not far from where Colton was drilling. Leroy Fried was working on Cyclone Drilling rig number 19 on a well owned by Ultra Petroleum. The screws on a support beam probably sheared through, loosing the beam that fell onto the rig floor before bouncing up and striking Leroy on the right arm and right leg (breaking both) and in the chest and stomach. It took the ambulance crew fifteen minutes to race from Pinedale to the well site, during which time Leroy's crew mates tried to keep him calm and control the bleeding.

"Just lie here nice and still."

"You're gonna be okay."

"We called the ambulance."

"Hang in there, you're okay."

And Leroy crying out for all of heaven and earth to help him.

"We're doing all we can. You're gonna be fine, man. You're gonna be fine."

The next few nights the talk in the bars around town and in the man camps all around the high plains circled around all the usual subjects: women (or the lack thereof), money, guns, pickup trucks—and the death of Leroy Fried.

"Poor bastard was still talking when they got to him," said one roughneck, "knew he wasn't gonna make it."

"He only worked for the company for sixteen days," said someone else.

"Poor fool."

Colton stared at his plate of meatloaf and wished it had been made by Melissa.

"Assholes don't give a shit what kind of rig they put you on," said another roughneck.

"Bolts were overstressed."

"Poor guy."

"They say he asked for his mom at the end."

"They say all guys ask for their mom before they die. Like if you get shot or what have you."

"That's just in movies."

"No man, it's for real."

"How'd you know?"

"I just know, man. It's one of the things you know if you're half-a-freakin'-wake, man."

"I'd ask for Pamela Anderson."

"Retard! It's not like you get who you ask for. It's not a request for a final meal or anything."

Colton pushed his tray away. "This meatloaf is horrible." He got up. "See you guys."

Someone pointed a fork at Colton's back. "S'up with Colt? Prickly as a porcupine."

"His wife's due any day."

"Oh."

• • •

Colton made his way down the sterile corridor of the trailer. Everything was temporary, cheap. There was a plant in the corner, a plant next to the door, a plant next to the bathroom door—all plastic. Kaylee always said never eat the meatloaf in a restaurant with plastic foliage or plastic knives and forks and Colton thinks, "Good freakin' advice, Mom." Next to the kitchen, there was a grey-carpeted room with a massive television screen. The boys were watching something blow up, fireballs, cars doing slow pirouettes in the air, metal gymnasts spinning around and around themselves. "You'd think they have enough of that all day," Colton thought. He went out into the mountain-cool summer night and made his way to the dormitory trailers where the beds were lined up in empty white cubicles. Colton took his greasers off at the door, sat on the edge of a bed made up like an army cot, and found his cell phone.

"Hello?"

"M'issa? It's me."

"Colt."

"How you doing?"

"I feel like a whale."

Colton smiled, "I bet you're beautiful."

"I can't tie my shoelaces."

"So? Wear slip-ons."

"It's any day now, Colt. You coming home?"

"Soon baby. Five more days."

Melissa sighed.

"How's Nate?"

"Missing you."

"Yeah."

"Yeah."

Colton scratched the back of his neck. "Baby?"

"Yeah."

"Nothin'."

"What?"

"Nothin'."

"Is everything okay there?"

Colton said, "Great. It's great. Couldn't be better. Getting kinda chilly at nights already."

There was a silence then Melissa asked, "Can't you get a job down here? It gets kinda chilly nights here too."

"Not that again, M'issa."

Melissa started crying.

"M'issa, hey M'issa. I'll be home soon."

"I miss you something fierce. Colton, please."

"It's good money."

"It ain't worth it."

"I'll buy me a F350. Tell me that ain't worth it."

"It ain't worth it."

"Aw, c'mon baby."

"Please come home."

"Five more sleeps."

Melissa sighed again and blew her nose.

"Cowboy up, cupcake," said Colton.

Silence.

"I love you, M'issa."

More silence.

"C'mon baby. I'm oil-field trash. It's what I do."

Melissa took a deep breath, "I love you too, Colton."

"That's my girl. You take care of our baby and Nate and I'll be home soon."

"Okay."

Colton ran his fingers under his eyes and held his face in his hands, the cell phone propped up in a couple of fingers to his ear. "I sure miss your meatloaf," he said. "Freakin' camp cook here makes the sonofabitchest meatloaf you ever ate."

"Promise me you'll be here for me," said Melissa.

"Yeah," said Colton. "Sure. Sure I'll be there. Of course I'll be there."

"Okay," said Melissa.

"Okay," said Colton.

Then the two of them sat on the receiving end of each other's breathing until Colton said, "I'm gonna hang up now and get some sleep. I gotta be back out on the patch in eight hours."

"Okay. I love you."

"Yeah, I love you too, baby."

Colton hung up and lay on top of the blankets with his hands behind his head staring at the white ceiling of the trailer. The place was starting to fill up with the sound of men breathing in an exhausted, industrial way. The air was rank with old sweat and testosterone overlaid with the sting of disinfectant. "Sonofa!" said Colton, turning on his side and shutting his eyes.

37

DAKOTA JUSTUS BRYANT

⸻ ❖ ⸻

September 15, 2004, Colton made it back to Evanston to see Dakota Justus Bryant being born, but he was halfway down the corridor on his way back to the Upper Green River Valley almost before the child had been rubbed dry.

"Where's he going?" Kaylee asked.

"Back to the rigs," said Melissa.

"What?" said Tabby.

Melissa started to cry.

"Hold on here, girl," Kaylee told Melissa. "He'll be right back."

"Me too," said Tabby.

Tabby and Kaylee ran out of the hospital room, down the corridor, and stopped Colton before he got to the sliding doors.

"Where do you think you're going?" said Kaylee.

"I've got a shift coming up."

Tabby poked Colton in the chest. "You go back into that room and sit by your wife's side. How can you be so ignorant?"

"I've got a shift coming up," shouted Colton.

"I don't care. She needs you," Tabby shouted back.

"What do you know?"

"I've had a baby. That's what I know!"

"Would you two stop yellin'?" said Kaylee. "We're in a hospital. There's people tryin' to be sick in here."

"Then tell him not to be so ignorant."

"Son," said Kaylee, "unless you'd like me to knock you flat on your ass, you're going to go back into that room with your wife and son. They need you."

"They need me to make money," said Colton.

"They need you here." Kaylee put her hands on Colton's shoulders. "And I'm tellin' you to go to them."

"Sonofabitch!" said Colton. He shrugged Kaylee's hands off.

"Watch your language, son."

Colton pressed his knuckles into his eyes. "Holy cow," he said. "If they fire me, then what?"

"Any company that doesn't understand that you need to be here for your wife and your children right now doesn't deserve your time," said Kaylee.

"Tony was there for me when Tanner was born," said Tabby.

"Yeah," said Colton.

"Your father was there when you were born," said Kaylee.

Tabby poked Colton in the ribs. "Yeah, Pant-leg Pete," she said. "Right there in the car with you."

Colton held up his hands. "Okay, Mom, I'll call camp and see if I can get someone to cover me for a few shifts."

"That's better," said Kaylee.

Tabby shook her head. "Sheesh, Colt. You can be so ignorant sometimes."

"Come here," said Colton, opening his arms.

Tabby put her head against Colton's chest.

"Sorry I yelled at you," said Colton.

"I'm sorry I called you ignorant."

"Okay," said Colton.

Tabby took Colton's hand. "C'mon, baby brother," she said.

Kaylee watched the two of them walk down the corridor back to Melissa's room, hand in hand, Tabby all bunched sideways so that her blonde ponytail touched the bottom of Colton's neck. Then Kaylee phoned Bill at the oil patch down in Utah. "Well," she told him, "I think your son's just had his first real taste of fatherhood."

"How's he taking it?"

"He's scared half to death."

"He'll get over it."

"I think he just did."

"Is the baby okay."

"Looks exactly like Colton."

"That a fact?"

"Peas in a pod."

"I thought we broke the mold."

"Me too."

"How's M'issa doing?"

"She's gonna be okay."

"I'll call you later."

"Tomorrow. I've got a late shift coming up."

"Bill?"

"Yeah?"

"Oh nothin'."

But she hung on for a moment and shut her eyes.

"Okay, girl," said Bill.

"Yeah, okay," said Kaylee and hung up.

38

Colton Quits

—————◦◦◦—————

Usually Colton drives half the night on his week off from old habit and also to keep from switching the graveyard clock on himself; a sixty-ounce bottle of Mountain Dew on the seat next to him, a tin of Copenhagen on the dash, a .22 across his lap. Then he'd phone Tony or Jake from somewhere down near Bitter Creek or Poison Basin and say, "Guess where I am?" or "I'm on a dirt track somewhere your side of Antelope Hills and I can't see the sonofabitch way to get me back to Evanston." And it would be three o'clock in the morning. "The only road sign I can see says, 'No.'"

"No what?"

"I dunno. The rest of the words are shot all to holy crap."

But recently he'd been staying close to home on his week off, spending time with Nathanial and Dakota mostly, and trying to get it figured out with Melissa. They'd been having some hard times these last months—his weeks on the drilling rigs in the Upper Green River Valley making lonelier and lonelier gaps in their time together and Melissa wanting him to quit the rigs worse than ever and Colton, never having been good at expressing himself, getting quieter and quieter before blowing up. "What else do you want me to do?" But what he really meant was, "What other choices do I have?"

Melissa tried to tell Colton that he could feel like free fall if you were on the receiving end of him. "You're always all or nothing, Colton. Why can't you do in between for a change?"

Colton let the question sink in for a moment, considered it seriously, then he bit his nails apologetically. "I never did find that gear," he said, which was a fact. So Melissa kept trying to press Colton into one place, the best way she knew how, taking him off to the photo studio every month or two for a family portrait and he, sullen in the resulting pictures, refusing to take off his ball cap, looking at the photographer like, "Just take the freakin' picture, would you?"

And then there were arguments. "Could you just smile at the man?"

And Colton replying that he didn't see the point to all these portraits, and he was never doing another one as long as he lived, so help him. And whatever else he did, he wasn't ever going hunting with this photographer. Did he not know where the trigger was on his damn camera? "I ain't gonna get any prettier, so just shoot it already."

• • •

But one morning in mid-August, just before Dakota's first birthday, Melissa comes out into the kitchen with the toddler on her hip and she says, "I ain't doing this, Colt."

Colton props himself up off the sofa. "Ain't doin' what, baby?"

"I can't do these two weeks on, two weeks off. I can't do the fighting about it. I can't do being scared half to death half of every month of every year. You need to quit. We got Dakota *and* Nathanial now."

"I already told you," said Colton, lying back down, "I'm oil-field trash, baby."

"No you ain't. You're a lot more besides."

"Maybe," says Colton. "But I'm nothing else that pays the bills."

"There's other work around here."

"Not good work."

Melissa levels her gaze at Colton. She puts Dakota down and

he crawls toward the sofa. "You go back to the rigs and you can stay there."

"What?"

"It's them or me."

"Holy crap," says Colton, sitting up again.

"Yup," says Melissa.

Dakota pulls himself up on the sofa. "Come here, Koda," says Colton, pulling the boy onto his lap. He wraps his arms around Dakota's shoulders and says, "Hey there, cowboy."

Dakota squirms to get down and starts to race, on all fours, for the door. Melissa scoops him up. "Which is it?" she says.

Colton looks at his hands.

"I don't want our boys following you out there. Three generations of Bryants on the oil patch is enough. I want our kids doing something different. I want 'em to go to college."

"Fine," says Colton, "but meantime someone's got to get the freakin' fuel for rocket scientists or whatever the hell it is you want them to be and that someone might as well be me."

"No," says Melissa.

"You're serious?"

"As a heart attack."

"Okay baby, I'll try it. I'll try to do ordinary construction or whatever the heck else they got out here for some boy ain't got no college degree."

Melissa smiles and puts the child down. "Okay," she says.

39

Colton Works
in Evanston

———————

But before hunting season was even over, not even two months into his new job, Colton came home from the toolsheds, threw his lunchbox on the counter, and handed the phone to Melissa. "Call my old boss and get me my job back on the rigs," he said.

Melissa stared at him.

"I said I'd give it a try working off the patch. I did and I can't do it."

"You barely gave it a try."

"It's enough for me to know it ain't my thing," said Colton.

Melissa lit a cigarette and squinted at Colton through the smoke.

"Get me back to the Upper Green," said Colton. "There's a rig with my name on it out there."

"That's what I am afraid of," said Melissa.

"I didn't mean it, that way. C'mon. Accidents happen all the time. My boss here is a freakin' drunk. I'm gonna get in some drunk drivin' accident with that fool. I'm safer out there on the rigs."

"No you ain't."

"It's overrated," said Colton. "It's one of the safest jobs out there. You're more likely to die in a car wreck. You're more likely to die of a heart attack. You're more likely to die of a dog bite."

. . .

But a week later it came out. On October 29, 2005, in the Upper Green River Valley, on a well site owned by Ultra Petroleum and drilled by Grey Wolf, Dewayne Hughes, a forty-four-year-old father of four from Casper, died when his safety harness became entangled in the rotary head of the top of the drill during a routine cleaning operation. For some minutes, until the rig could be shut down, Dewayne spun with the drill at 45 rpm, inexorably into the earth, the harness pulled so tight against his body that he suffocated. He was four shifts into his third hitch on the oil rig, he had not any previous oil-patch experience, and he had not completed any safety training. Other members of the crew were new to the rig too. The floor hand had worked on the rig for three hitches, the motor hand had been two days on the rig, and the derrick hand had been on the rigs for three months.

That week's newspaper was thrown on the top of the coffee table. The death made a small impression on the front page, a couple of inches of type under the heading, "Hand Killed on Ultra Rig." Colton snatched the paper up and laid it down on the carpet, exposing the real estate pages to the ceiling, "I got to clean my gun," he said. "It's hunting season."

Melissa lit a cigarette and sat down on the sofa. She held a can of Mountain Dew on her lap.

"Well," said Colton, "my dad always told me there's just a couple of things a man has to take care of in his life; his gun and his wife . . . Well, I guess more than a couple, 'cos then there's his kids and his nieces and nephews and his horse, his pickup truck and his boots and his saddle, his sisters, his mom. I mean, there's a whole heap o' things a man has to take care of, but you can tell a lot about a man by how he takes care of his gun."

"And his wife," said Melissa.

"Right," said Colton. "And everything else. Man, it's no

wonder my dad's so thin. You burn off your feed just thinkin' about it."

Melissa smiled.

Colton called Nathanial over to the newspaper. He broke his gun into pieces and lay it out for the boy to see. Then he got his gun oil and some mutton cloth and gave a piece of clean cloth to his son. "Okay kid," he said. "Ain't no point havin' a gun if it ain't taken care of and the same applies to everything else in this life."

Nathanial sat up on his knees and his face went all tight and serious.

"Difference is," said Colton looking down the broken barrel of his gun, "between you and me, you just have to take care of your mom and your baby brother. Me, I got the whole darned planet on my shoulders."

Melissa smiled again and shook her head, blowing smoke at Colton. "He's just being a drama cowboy as usual," she told her son. "Truth be told, it's the women who carry the weight of the world."

"Wanna fight about it?" said Colton.

"Only if you fetch me another Mountain Dew."

Colton laughed, "He-he-he."

40

MINUS THIRTY-FIVE

It's been a cold month or two in the Upper Green River Valley since Colton's been back on the rigs. Down to minus thirty-five and the conditions are enough to break a man's soul, let alone a steel cable or a metal bolt. And still the rigs are expected to bore down into the hardened earth, relentlessly. Colton is back around Evanston from a hitch in the Upper Green River Valley for six days. He and Tony take a day to drive to Rock Springs for Christmas shopping.

Colton says, "Another roughneck on our patch just got nailed out there this week."

"Another one?"

"Yeah."

This latest one—a drill hand on a Cyclone rig for Ultra Petroleum—had been doing it twenty-five years. But they say the fast line sheave was wore out like a piece of dental floss. They say the traveling block, hook, and elevators—the whole lot—all came crashing down onto the floor and that the whole rig shook for fifteen minutes afterward. They say the drill hand was one of those who liked to do a little meth to keep awake and then a little pot to smooth out the high and on top of that the rig was falling to pieces. That'll be enough to do 'er and it was bad luck anyway. All the other hands made it into the doghouse, except the drill hand, who got caught with one leg in the door. He was sliced open from the base of his neck to the seat of his pants and the tool hand went into shock on the spot because the guy died right at his feet, and he had already lost his only brother, killed on a rig.

"Holy crap," says Colton, "it's been cold out there."

"Yep," says Tony.

"I think it's gonna be the death of me," says Colton.

"The cold?"

"No, the rigs," says Colton. He stuffs a wad of chew into his lip. He looks out the window at the way the world is flat-lit in dreary winter grey. The wind is making scallops out of the lie of the land, coating fences and trees with a fragile blanket of rime, threading ice into the heart of the world. There's a herd of horses as still as crusted statues on the desert. On days like this, the ground is nothing more than the impression of heaviness beneath your feet, the sound of your tires gripping slick nothing. "Sonofa," says Colton and spits into an empty Mountain Dew bottle. All the radio stations have nothing on but Christmas tunes, like a nervous complaint.

Tony puts in Sara Evans. "That's better," says Colton. "Freakin' 'Jingle Bells' gonna be the death of me."

Tony laughs and looks at his brother-in-law. "What *isn't* gonna be the death of you, Colt?"

Colton spits. "He-he-he," he says.

PART TWO

41

THE DAY BEFORE
VALENTINE'S DAY

Evanston, Wyoming

———— • ————

The day before Valentine's Day 2006 near Evanston, Wyoming, the sky had come down to meet the earth in such a way that everything was a sameness of grey-white and boundaries were nowhere. Snow blew sideways, out of the northwest, so that it hissed against the siding. Colton heard the weather before he moved. He pictured horses out there, backs to the wind, tails under bellies. He knew without needing to see it, the way trees would be winter-brittle barely holding on to the earth, the way freeze-dried livestock would be pressed up against the bruise of willows in the creek bottom. He understood the ridges up above town would be scoured bald, snow veined thinly between tufts of tough grass. Colton swung his legs out of bed and scratched the back of his neck, already thinking of the drive north to the Upper Green River Valley on roads like this.

He went through to the kitchen, cracked a Mountain Dew, and drank it like it saved lives, folded a piece of bread over a slice of cold ham and swallowed it down with the rest of the soda. The clock on the microwave said not yet seven. It was still night out there. Colton zipped himself into coveralls, pushed his feet into his work boots. He pulled a baseball cap onto his head—black with the words western petroleum in orange-and-white letters across the front, the peak tunneled just so, tight against his head, so it couldn't blow off. He put on his fleece-lined work gloves and opened the door to outside. The wind caught him on the face like the flat side of a two-by-four and took his breath out of his lungs. It was weather with a violence, like hell had been ordered in, everything on the edge of its frozen limit. He stamped his way out into the dark, snowflakes landing on the porch like Styrofoam bubbles, too frozen to be sticky.

The neighborhood was starting to take shape against the dawn. The playset Colton had set up for Nathanial and Dakota swinging crazy in the storm like ghost children maddened with the endlessness of being dead; the dog kennels over the road crouched low against the weather like igloos; the horses' shelter next door singing against gravity, the tires on its roofing sheets the only thing between them and cartwheeling free, and the neighbors' dogs were shouting *wa-wah-wah-wah*, as if it would make any difference to say the same thing over and over.

Colton seared this place into memory, loving it. Then he went inside, peeled out of his work clothes, and made bacon and pancakes. At seven-thirty Melissa came out of the bedroom with the kids, one on each hip. She put the kids down. Dakota was just learning to run in that bowlegged sailor way of babies, leaning against the tilt of the earth's orbit. Nathanial ran up to Colton. "Pancakes!" he shouted, excited.

"Why the pancakes?" asked Melissa, lighting a cigarette.

"Tomorrow's Valentine's Day," said Colton.

She frowned and blew smoke into the kitchen. "I've got to get to work," she said.

. . .

Colton took the boys to daycare at nine, Dakota curled up in the crook of his elbow, not wanting to let go. "Hey boy," said Colton. "Koda?" He put the child down. "Hey, son." He held the child's chin in the vee between thumb and forefinger. "You take her easy, you hear?" And then to Nathanial: "You're the man of the house while I'm gone, okay?"

Nathanial wrapped himself around Colton's leg. "Can I come with you? I want to come with you, Daddy."

"Where's the hair on your chin, son?"

"I got hair on my chin! I got hair on my chin!" Nathanial started to cry.

"Hey Nate-ate, cowboy up now." Colton felt tears coming. He cleared his throat. "Nate, I got to go." He unfolded and walked back out into the sideways snow, hand up alongside his cheek against the sting of the wind, stamping, as if angrily.

• • •

Colton went home and took a nap for a few hours, powering up for the drive north and a two-week shift of all night on the rig. When he woke up, he lay for a few moments in the grey-dark room and paid attention to his gut feeling, the way another man might check his body for aches and pains. Then he sat up and took off his wedding ring. He put it on Melissa's bedside table along with his wallet, peeling out a fifty that he stuffed into his pocket.

Half an hour later Colton was on his way driving north in his white Ford F150. There was an empty gun rack behind his head, money stashed between the seats, between the console and the seat, behind the visor, in the glove box—his flat-tire money. The Nitty Gritty Dirt Band was on the player, but Colton's mind was not on the music. Not seven miles into the trip, just this side of Carter, he pulled a U-turn in the road, and drove an hour out of his way, back to Evanston to reckon with Melissa where she worked—a stuffed dog, a dozen long-stemmed roses, and a box of chocolates from Wal-Mart. "I love you," he told her. "We'll get her figured out. Okay?"

Melissa's shoulders went soft. She smiled, holding the stuffed animal around its neck. "Go on," she said. "You'll be late."

"You'll bring us some meatloaf to the man camp?"

"Of course."

"And when I get back we'll go fishin' in the dark."

"Oh, get out with you." She pressed her lips together and looked away.

"I'm gone," said Colton with a cockeyed grin, and he half danced, half bounced back out into the weather, shaking his hands up by his ears like he was getting rid of the nuisance that was himself.

"Be careful out there," said Melissa.

Colton got back onto the road and hoped nothing would be blown closed. Out east, on I-80 between Cheyenne and Laramie, there were news reports that tractor-trailers had been flipped onto their sides, SUVs trapped on the highway were being blown across the ice from a standstill. Colton picked an imaginary line for the edge of the road, set himself steady at fifty, and in this way he drove north all the way to Big Piney.

42

CUMBERLAND CEMETERY

Between Evanston and Kemmerer, there are windmills on Highway 189 that follow the traffic here, waving ponderously, and then they disappear off the face of the earth, their slow-swinging arms like dinosaurs engaged in some thoughtful farewell dance. Then after that, there's nothing but snow-blown hills and strands of barbed wire and a railway line until a BP sign that says private property violators will be prosecuted and behind the sign the whole of the rest of the United States from here to the Pacific Ocean.

Within a few hundred feet of the BP notice there's an old white picket fence with a gate, and above that a sign saying CUMBERLAND CEMETERY in metal letters against the sky. The cemetery is the burial grounds for two old coal-mining towns originally called Little Muddy but later renamed Cumberland I and Cumberland II. Many of the miners in Cumberland were survivors of the Ludlow massacre of April 20, 1914, when two women, twelve children, and six miners were gunned down in their tent colony at Ludlow, Colorado, by the Colorado National Guard in response to the Colorado Coal Strike. So they came north for a better life and all that's left of Cumberland I and II and that awful ache of bad luck that the miners brought with them is this little graveyard.

Most of the graves inside the tiny cemetery have small white wooden crosses, and whatever names they possess have been worn off by wind and weather, knocked sideways by cattle and sheep.

Some of the scoured headstones have some names you can read, though, etched in what looks like a child's unschooled hand. And then the heartsore realization comes that these are all children's graves, barely a few months old many of them, pressed into the earth at the turn of the first quarter of the last century, some of the dates on the stones reversible, as if the children came to earth with one foot still in the spirit world so that the journey back would be all too easy:

> *Lovean Wilde 1926–1926*
> *Baby Anderson Nov 16,1927–Nov 16,1927*
> *Grace Blacker 1915–1915*
> *Baby Son of Geo. & Mary Blacker*
> *Bernice Tremelling 1915–1917*
> *John G. Faddis Oct 1909–Dec 1909*
> *Ferrell Wilde Jr. 1924–1924*
> *William Blacker 1900–1917*
> *Henry T. Blacker Oct 18,1910–May 13,1913*
> *Mae Tremelling Jun 28,1914–Jun 19,1916*
> *Daniel McWilliams 1913–1913*
> *Thelma D. Patterson Feb 19,1904–Feb 21, 1906*

And still the graves pile tiny hump next to tiny hump. Did the wind snatch all these children up? Drought? Influenza? Poisoned air or water? There is no other sign that anyone lived here once. There is nothing like a church, a bar, or even the foundations of a church or bar. There is no sign that there were owners of these children. And beyond the cemetery fence nothing but ancient slabs of cow manure, rock-hard pebbles from sheep, and a high, lonely wind off the ridge. The sound underfoot is like the pitiless crunch you hear on old made-for-television Westerns—either crusty snow or flinty soil, nothing soft or lullaby here.

. . .

Just beyond the Cumberland Cemetery, something made Colton think to call Jake. Maybe it was all those names on all those wind-worn headstones with no one to remember who was lost. Colton kept an eye on the road, one hand on the wheel, and waved his cell phone around to see how many bars he had. He hit speed-dial two.

"So how does this work?" he had asked Jake once. "I mean, if George W. Bush presses speed-dial two on his cell phone does he get you too?"

"No, Colt, he gets Laura. Or Cheney, probably."

"Why don't I get Cheney?"

"I hope you're kidding."

Colton made that face.

"You're such a freakin' redneck, you know that?" said Jake.

"You might be a redneck if," Colton said, "someone asks to see your ID and you show them your belt buckle. You might be a redneck if . . ."

Jake knew this could go on all day. He walked away with his hands over his ears. "I'm not listening to your nonsense, Colt. You're enough to turn a good man bad."

"You might be a redneck if you think ketchup is a vegetable," shouted Colton.

"Like the man said, Colt, I'm not listening to you anymore."

. . .

It was just past four and Jake was driving home from Pinedale to Boulder. If you'd asked him there and then he would have said that, on the whole, things were going surprisingly easily. He was only twenty-three years old, making his way in the oil business. Tonya didn't have to work, so their two young children were taken care of at home—the way kids should be taken care of. Jake liked his job, he had the weekends off, and worked a regular eight- or nine-hour day instead of those twelve-, eighteen-, twenty-four-hour shifts flow testing, and, as he said, "Not bad for a kid they all said wouldn't amount to much."

Jake's cell phone rang. Here it was, already dark with that northern winter's finality, and the snow making crazy tunnels of speed, flying at him along the beams of his headlights, fresh white coming at him from all sides, even, apparently, from some unseen place underground. Jake said, "I'm not answering in this weather," but then he saw who it was and answered anyway, shouting because reception was always touch-and-go just here. "Colt?"

"You bolted down?"

"No, I'm driving. Where are you?"

"This side of Kemmerer. I got a hitch coming up."

"How's the road down there?"

"Nothing I ain't seen before."

Jake shook his head. "Man, this is, like, *freezing* to the other side of the earth."

Colt laughed. "Hey, it's big, beautiful Wyoming."

"Yeah, well it sure ain't Florida."

"How's the family?"

"Good. Yours?"

"Good."

"Hey, I got to hang up and drive. It's wild out here."

"Keep it on the road."

"You too."

"Hey Jake?"

"Yeah?"

But there was silence, that kind of drop-dead silence of a cell phone that's been bumped out of its range.

Jake said, "Colt? Colt, you there?" He looked at the screen on his phone. He'd hit the dead spot by the New Fork River. "Freakin' backwoods," he told the Upper Green River Valley.

43

VALENTINE'S EVENING

Jake and Tonya

It's a perfect house for Jake and Tonya, on the north side of the road from the oil patch—if it weren't for the way the high plains start off as a bench, they'd be able to see Colton's red, white, and blue rig from their sitting room. Out back where they plan to build a porch one day, there are views of the Wind River Mountains. There's a big yard for the kids, a few acres of paddock for the horses, a kennel for Jake's retrievers, and a chicken coop for the hens. Real estate prices in the Upper Green River Valley have soared since the beginning of this boom. A room in someone's basement will cost you a thousand dollars a month, the mortgage on a three-bedroom house would set an ordinary man on his ass. But Jake can afford for things to be a bit pricey. And the way he figures it, the cost of housing keeps the riffraff out. Well, except for the meth heads next door with their crummy collection of trailers and broken-down caravans and junked cars, the whole heart-rotted property marked with signs that say private, keep out, and beware of the dog.

· · ·

"Sure you got all the air out of that can?" says Tonya.

"Yeah," says Jake.

Tonya frowns. "You okay?"

"Yeah, I'm okay."

Tonight Jake and Tonya are up after they've put the kids to bed, trying to add to the three months' store of food they are supposed to have set aside for the end of the world as we know it; "if ye are prepared ye shall not fear." Members of the Church of Jesus Christ of Latter-day Saints have published a whole body of literature on the science and philosophy of food storage. The index of prophetic statements regarding possible plagues for which such storage should be prepared reads like a bad evening of news: war; economic collapse; imminent nuclear fallout; famine.

A surplus of wheat and out-of-season produce on sale a week earlier at Faler's General Store on the outskirts of Pinedale—"All the Civilization You Need"—have prompted Tonya to stock up and get up to date with her disaster preparedness. She'd been boiling and peeling all afternoon and the kitchen has a metallic, blood smell to it. Tears of condensation run down the walls by the stove. The Valentine's roses Jake has bought her are already hanging their heads, as if in tropical exhaustion.

But Jake doesn't have his heart into it tonight—neither the end of the world, nor the canning—and he's ticking Tonya off with his attitude. He keeps going to the window.

"What's wrong with you?"

"Nothing. What's wrong with you?"

Tonya shakes her head. "Holy cow," she says. "Hand me those tomatoes."

"They're not ready."

"What does a tomato have to do to be ready for the dark?"

"Do you want exploding tomatoes?"

"Would you just hand them to me? I know what I'm doing."

"Here," says Jake.

"You're making me edgy," says Tonya.

Jake goes to the window again. "At least it's stopped trying to pull the rigs out of the ground," he says.

"Is it still snowing?"

"Nope."

And then, at 9:45 p.m., there was a sound that came through the walls of the house, like the wind picking up again, a high howling noise, but it rose and fell so that Tonya puts down the jar she is labeling and she wipes her hands on the apron. "Ambulance?" she asks. "Cops? What's going on?"

She goes to the window and stands next to Jake, both of them with their hands cupped against the pane. They watch police cars coming out of the pale nothingness between Pinedale and here, ghostly creatures streaking with their urgent blue/red/white lights in panic mode against the white-winter night.

"Maybe they've come to get the crackheads," says Jake.

"Dream on," says Tonya. The cop cars sail right on past the subdivision with Jake and Tonya's house and the little methamphetamine triangle next door and keep going out onto the desert. "See, I told you."

Then an ambulance, flashing red like a heartbeat monitor onto the white end of the snowstorm, comes rushing past as best it can on the snow-packed roads, following the cops. Then another police car wailing down the lonely white road.

"Must have been an accident on the rigs," says Jake.

"Must be," says Tonya.

"Holy crap," says Jake. "I hate that."

"C'mon," says Tonya. "You've got an early start. Let's finish up and get to bed."

They are cleaning up the kitchen when they hear the helicopter.

Jake looks up. "Somebody got it bad," he says.

"Flying on a night like this. Must be serious."

"I told you." Jake scratches the back of his neck. "I told you," he says.

"Told me what?"

"I dunno. Nothing."

44

FREE FALL

By nine o'clock the storm had taken a corner in the mountains and there was a pause in the weather. The plains looked refreshed by the recent snow, moon-glowing under the winter sky. It was ten degrees, the wind barely thinking about it. The men started the shift by taking some drilling survey measurements and Colton was sent under the drilling floor to find a power source for the survey equipment but he came back up to find that the men were looking for a twenty-four-inch wrench to attach the survey equipment. At that moment the well was fresh, only 324 feet deep.

At around 9:10 Colton was walking around the catwalk for at least the second time in a space of ten or twenty minutes. It has been supposed that he was going to get the wrench. As usual, there should have been rails. As usual, Colton should have been wearing a harness. As usual, there weren't and he wasn't and this isn't recreational exposure. So here's Colton without a chance to catch himself, slipping through the mouse hole into the cellar, and who knows when the blow comes but at that moment, the great plains become a dark sea and everything that Colton was, is swallowed up in its waves.

His hard hat and the twenty-four-inch wrench land next to his body.

It's nearly nine-thirty by the time the floor hand sees Colton down there, but the light is terrible and the noise from the drilling is drowning out his voice. "Colton!"

Colton is lying like a comma, a pause, against the cellar floor.

"Colton! What you doing down there?"

Colton doesn't respond.

"Holy crap," says the floor hand and hurries up to the drilling floor. "Colton's hurt! I think Colton's hurt."

The drill hand looks at his watch. "Oh shit," he says. "What'd he do?"

"Man, he ain't moving, he's in the cellar."

"How'd he get there?"

"By the looks of it, he fell. He ain't moving."

"He ain't moving?"

"He ain't responding at all."

"Oh shit, oh shit," says the drill hand and he's almost in free fall himself, running down the stairs to the rig and then below that into the rig's cellar where the cold is metallic in its density.

45

JAKE DRIVING ALL DAY

Jake woke up at three the next morning to get to work. He took a quick shower and looked out the window. The storm from last night had come back with fat, quick flakes, a driving snow, in a hurry to cover itself up. Then the phone rang and Jake said, "What the heck? Who can that be at this freakin' hour?" He picked the receiver up, "Hello?"

"Jake?"

"Yep?"

"Jake, it's Shad."

"Who?"

"Merinda Bryant's boyfriend."

Jake felt the blood rushing from his neck. "What's happened? Who's been hurt?" he asked.

"It's Colton."

"Oh no."

"He fell off a rig last night."

"Holy crap," said Jake. "Oh please God, holy crap. Is it . . . Is it?"

"He's in the hospital here in Salt Lake. They life-flighted him down last night."

Jake felt the floor rush away from him and his knees buckled to catch up with it.

"Jake?"

"Yeah," said Jake, "I heard the chopper."

"It's not good, Jake."

"I guess not," said Jake. He held the receiver away from his mouth and retched.

"The family felt you would want to be here."

Jake wiped his mouth with the back of his hand.

"Jake? Jake, you there?"

"Yeah, yeah. I'm here. I'll be there."

"He's hanging on but not for much longer."

"I'll be right there." Jake crawled to the bathroom and held himself over the toilet for a few minutes.

. . .

They drove straight through, from about three-thirty in the morning until noon. The snow was piling down so fast that the road kept shifting shapes under their wheels, but just this side of Big Piney they got behind a snowplow spraying up a tunnel of road behind the crashing rasp of its blade.

"Holy crap," said Jake. "Thank you, Heavenly Father."

Tonya closed her eyes. In the back, the children slept in their car seats, wrapped up in blankets and tucked under pillows. Jake was crouched up over the steering wheel, his eyes fixed on the rear lights of the plow. In this way, they make it through La Barge and past Names Hill where just about everyone from the plains Indians to early explorers to roughnecks from this latest boom have stopped on their way through here to engrave pictures or their names on the red cliffs on the west side of the Green River.

They followed the snowplow through Kemmerer and by now it was getting light. "Okay," said Jake. "We're gonna be okay."

Tonya said, "You want me to drive?"

"I'm fine," said Jake. And then he started crying. "Holy freakin' cow."

"He's gonna make it," said Tonya.

"Holy Father," said Jake, "please don't take him yet. Please, oh holy cow, please."

46

PATTERSON-UTI DRILLING

By midmorning the safety officer from Patterson-UTI had arrived at the hospital in Salt Lake City. Kaylee, Bill, Tabby, Tony, Merinda, Shad, Preston, his wife Mandi, and Melissa were all sitting in the waiting room next to the intensive care unit. They had not been allowed to see Colton yet—the doctor had said that they needed to try and stabilize the boy first. The family looked up when the stranger walked in. He introduced himself as the Patterson safety hand and then he said, "I can't believe it. I've been up twenty-four hours with another accident and now I got to deal with this bullshit."

Bill unfolded himself and stood between the safety hand and Kaylee.

"I tell you what," said the safety hand. "I'm gonna take a nap in the car and when I come back we can discuss this situation."

Bill nodded.

The safety officer walked out.

Tony looked at Shad. "Did you hear that?" he said.

"What the heck?" Shad said.

Tabby started crying.

Tony put his arm over her shoulder. "Take it easy, baby. Take it easy."

• • •

An hour later the safety officer returned. "Okay," he said, "I've talked to the bosses and here's what the company's gonna do for you. The company's gonna get a hotel room for you. If the boy dies, we can help with the funeral, but we got to get blood and urine outta him and test for drugs. He comes up hot for anything and you ain't getting nothing."

Melissa looked up.

"It's okay, girl," said Bill.

"Well then," said the safety hand.

"He's not dead," Melissa cried. "Why are you talking about funerals?"

Bill put his hand on Melissa's shoulder. "It's okay, girl," he said, "we'll take care of it."

Melissa shook her head. "Colton's not dead!"

"Boys," said Bill, looking around the room at Preston, Tony, and Shad. The men rose to their feet. Bill nodded at the safety officer. "I think we'd better discuss this outside," he said.

The safety hand looked from Bill to the boys and back to Bill. "Fair enough."

"If you see the doctor," said Bill to Kaylee, "send him down to the lobby."

The four men followed Bill out of the waiting room, into the pastel-wallpapered halls of the hospital, their footfalls muted by grey carpet tile squares. Down in the lobby, Bill squared himself in front of the safety hand. He leaned back on the heels of his cowboy boots and crossed his arms. "That boy," he said, "is in about the roughest shape a body can be in. He ain't breathin' alone, he ain't pissin' alone, and he sure as heck don't have clean blood right now. He's hot for morphine and stimulants and I don't even know what else. Now if you take a piss test and scrape some blood outta what little is left in his veins he's gonna fail the test, isn't he?"

"Well," said the safety officer, "then I guess he won't get worker's comp."

Tony flipped open a notebook and started to write.

"And another thing," said Bill. "As his wife says, our boy ain't

THE LEGEND OF COLTON H. BRYANT

dead yet so I'd appreciate it if you could refrain from mentioning funerals."

The safety hand passed a nervous hand over his lips.

Then there was a long silence, uncomfortable for everyone perhaps except Bill, who had always found silence the easiest place to be. The safety officer glanced at the door once or twice and every time he did Bill shifted slightly.

Eventually the doctor came down. "I understand there was some request for a drug test," he said.

"That's right," said the safety hand.

"I specifically do not grant you the permission to do any such thing," he said.

Bill nodded and uncrossed his arms. "Thanks, Doc," he said. He looked at Preston, Shad, and Tony. "Okay boys, that's all." He nodded, turned on his heels, and made his way back to the waiting room.

47

TOUGH ANGEL

———◆———

By noon Jake and Tonya had arrived with Jake's parents. The doctor talked to the family and told them that Colton was as comfortable as he could ever be, but that there was nothing more that anyone could do for him. He'd never breathe on his own again. He'd never walk or talk again.

"What does that mean?" said Melissa. "Does that mean he's . . .?"

The doctor sighed. "I'm sorry," he said, and shook his head.

Melissa sank onto the floor and covered her head with her hands. "Oh my God," she whispered. "Oh my God."

Shortly after that, the nurse came out into the waiting room and sat in one of the empty chairs. Tabby noticed she was wearing a fresh tunic, but there was blood on the cuff of her trousers. "I want to warn you," she said, "he's in pretty bad shape. You need to prepare yourself. When he fell off the rig, he must have hit his head on something on the way down. There's a hole above his eye." The nurse swallowed and shook her head. "I'm so sorry. It's a"—she made a fist and showed the family—"it's a big hole." She took a deep breath. "And he's pretty beaten up. I mean he's black and blue and swollen. You just need to know this before you see him."

Kaylee nodded. "We understand," she said.

Bill put his hand on Melissa's arm and helped her to her feet. "Thank you, ma'am," he said. "We're ready to see him now." He looked around at his family and barely nodded his head.

. . .

Colton was lying in a small white room on a bed, a tangle of wires and tubes coming out from under the sheets that were pulled up to his chin. His face was swollen and above his left eye a light gauze dressing covered a deep purple hole, like something that a body sustains on the battlefield. His chest rose and fell in time with the mechanical rasp of the respirator. His family stood around him in stunned silence. Nothing the nurse had said could have prepared them for the extent of Colton's injuries.

Bill cleared his throat and nodded to Jake. "Would you perform the healing blessing, Jake?"

Jake and his father stood on either side of Colton and laid their hands on him. Jake was taking shallow, nervous breaths.

"Go on, son," said Jake's father.

Jake dipped his finger in the blessed olive oil. "Colton H. Bryant," he said. "In the name of Jesus and by the authority of the Holy Order of the Melchizedek Priesthood I lay my hands upon your head and anoint you with this oil that has been consecrated in the name of Jesus Christ. Amen."

Jake looked at his father. Then he looked at Colton's family. He shook his head. "It's not the healing blessing he wants," he said quietly. "It's the final blessing."

Bill nodded. "I know," he said.

Kaylee started to cry.

"Oh my God," said Tabby, "no."

"He can't be like this forever," said Jake. "You know as well as I do."

"I know," said Kaylee.

Jake said, "It's what he would have wanted."

Merinda said, "I just want to open his eyes one last time then." She walked up to Colton and gently opened his eyelids. "I know I'll never see eyes so blue again," she said.

"He doesn't want to be around if he can't even breathe on his own," said Jake.

The nurse said, "You'll need to leave the room then, when we . . . when we . . ."

"Thank you," Kaylee said. She took a deep breath. "We understand." She kissed Colton's cheek. "I love you, son," she said. "God knows, I love you. We need to leave for a minute, but we'll be right back. You're gonna be okay."

Bill bent over too, but whatever he said to his son, he said so softly no one else could hear. Then he took Kaylee's hand. "Come on, girl," he said.

Everyone followed Bill and Kaylee out the door and into the waiting room where they waited in silence, as if their own breaths had been stopped.

Then the nurse came back into the waiting room. "You can go in," she said. "He'll slip right away now."

Merinda was first to the door. When she saw Colton lying there all alone, his breathing stopped in the awful finality of the respirator's silence, she cried out, "He's all alone in here! We got to be with him. He's all alone!" She ran to Colton's side and took his arm and Kaylee took a hand and Bill took the other hand. "Everybody hold him," Merinda said. "Hold him." Preston held one foot and Tony took the other, Jake held one leg and Shad held another and everyone waited. "Keep holding him," Merinda sobbed.

But Colton's heart kept going, *Da-dum, da-dum, da-dum.*

"Say something," said Melissa. "Say a prayer. Jake, say something."

"Father," said Jake, "take thy servant into thy hands and take care of him. Know that he was a good man."

Still Colton's heart kept beating.

"I brought him into this world," thought Kaylee, "and I've got to be here to see him out of it. Oh God, please give me the strength to watch him die." She stroked Colton's hand. "Son," she said, "I've got you. I'm watching. I'm holding you with all my heart. It's okay, Colt. It's gonna be okay. You did good. You did so good. You made me so proud . . ."

Bill squeezed Colton's arm and he thought, "What she just said, son. It's true."

"Colton, you're in good hands now," said Jake.

"It's okay," Merinda told Colton, "we're all here. You're okay. You can go when you're ready."

Da-dum. Da. Colton's heart stopped. Then it started again, *da-dum, da-dum, da-dum,* racing. For twenty minutes Colton's heart kept beating. Sometimes it would slow down, even stop for a moment, and then it would come back in a panic. "Colton," Jake said, "I promise you that for every breath in my lungs and every beat of my heart, I'll take care of your family as best I can. You gotta let go now. Your time here on earth is done."

Still Colton's heart fought on.

"We love you," said Bill. He leaned over and kissed Colton and said loud enough for everyone to hear, "I love you, son."

Colton's heart took one last run at it, *Da-dum, da-dum. Da-dum. Da-dum. Da. Dum. Da.*

And then there really was silence.

• • •

No one said anything for a long time. And then Kaylee looked up at the ceiling and smiled through her tears, "Good luck, God," she said.

Bill looked at Colton's face, the hurt and the fight all quieted out of it now. "You said it, girl," he said. "He's gonna be one tough angel."

48

Rainbow

Upper Green River Valley

———◦———

Colton H. Bryant was pronounced dead at 2:50 p.m. on February 15, 2006. For a long time after everyone else had left the room, Jake sat at Colton's side holding his friend's hand. He cried and cried some more until his eyes were almost swollen shut. After that, the nurse came and stood next to the bed and it was clear to Jake that he needed to let go of Colton's hand now. "Holy cow," said Jake, looking at the nurse, "he was my best friend."

The nurse said, "I'm so sorry."

Jake scrubbed his eyes with his fists.

"You take as long as you need," said the nurse. "I can wait."

Jake nodded. "I just got to do one more thing," he said, "and then I'll go."

"It's okay. You take your time."

"Thank you, ma'am," said Jake. He cleared his throat. "This won't take long," he said. "I don't even know all the words." He took a deep, shaky breath and then he started to sing, "If I should die before I wake, feed Jake. He was a good dog, my best friend throughout it all . . ." And then he hummed all the words he didn't know and then Jake told Colton, "Okay, I'm leaving now. You're gonna be okay. I've got to go." After that, Jake walked down the corridor greasy with the smell of hospital lunch and back out into the world through the glass doors and into the parking lot. And there was the world, going about its business as if nothing at all had happened and luck and love were on the side of all God-fearing boys in blue jeans. None of which would ever be true again.

Jake, Tonya, and the kids left Salt Lake City an hour later. The storm had completely cleared by now, leaving nothing behind it but a blanket of white beyond the melted, grit-crusted roads. Such a stillness had settled over the land, it was as if the whole world had cried and blown and snowed itself into still exhaustion. Jake made it to Evanston before five o'clock. He only had one stop there.

"Where we going?" asked Tonya.

Jake said, "I just gotta make sure of something."

He pulled into Front Street and stopped in front of Uinta Pawn. "I won't be long," he told Tonya. He went inside. The door made a cheap electronic bing-bonging noise. The place smelt of Colton's boots, it was true. "Excuse me, sir," said Jake to the man behind the counter. "I was wondering. Do you have anything in here belonging to Colton Bryant?"

"I'll need to see a ticket," said the man.

Jake pressed his lips together and looked above the man's head at a mount of a four-point buck.

"No ticket, no deal," said the man, turning away.

Jake said, "Colton's dead, sir."

There was a silence.

A tear rolled down Jake's cheek. "Can you just give me his stuff?" he said. "I'll pay."

"So help me," said the man, hurrying to the back of the shop. "I know you, you're always paying his ticket for him, ain't you? I know you." He came back with Colton's custom-made saddle, the one with colton in fancy western lettering on the back of its seat. He held the saddle out to Jake.

"He fell off a rig in the Upper Green last night. He died this afternoon."

"Oh man. I sure am sorry to hear that."

"What do I owe you?" said Jake.

"Not a cent," said the man. "Please, just take it."

Jake nodded and took the saddle.

"I just want to say . . ." said the man.

"Yes?"

"I just want to say, my condolences, sir. That boy was something else."

Jake nodded and walked back out into the silent world with the saddle, climbed into the pickup, and got back on the road north.

· · ·

Even the wind had stopped blowing, here, where the wind never ceases and the great, ponderous windmills on the ridge above I-89 looked frozen, they were so still. After an hour, Tonya slept in the passenger seat with Jake's coat pulled up to her chin, the kids slept in their car seats in the back, and Jake drove on and on, past the Cumberland Cemetery to Kemmerer and from there through La Barge and Big Piney and Pinedale and finally to their little house across the road from the mesa, just a mile or two as the crow flies from rig 455.

It was dark by the time they got home. Tonya put the kids to bed and Jake took a shower and went and lay down on the sofa. When she came back from the children's rooms Jake was staring up at the ceiling.

"What if it's all a bunch of complete BS," he asked, "and there's no God and no heaven? What if there's no nothin' after this?"

"Don't say that," said Tonya.

"Well? What if?"

"I'll bring you some hot milk," said Tonya. "Maybe you'll sleep."

"I ain't gonna sleep," said Jake, and then tears overcame him and he couldn't speak.

"Okay," said Tonya, "whatever you need to do." She fetched some milk from the kitchen and turned out the lights.

Against all odds, Jake slept deeply until dawn, although his dreams were disturbed and fraught and several times in the night Tonya heard him call out. He awoke stiff and surprised to find himself heartbroken. And then it came back to him, what had happened. "Holy cow," he said. "I hate that." He sat up and rubbed his head and then he got out of bed and went into the living room where the window looked across the mesa. What he saw next made him shout. "Tonya! You got to come here. Tonya!"

Tonya hurried over.

"Look," said Jake.

Over the calm sheet of snow that lay beyond the head of the trashed-out trailers in the junkyard next door and beyond the road, right above rig 455, there was a steady, bright pillar of the end of a rainbow, and not a cloud in the cold, clear winter sky.

"I've got to get pictures of this," said Jake. "That's exactly where he fell. That's most definitely Colton." He grabbed the camera, put on some boots, and walked out to the back of the house. He took several pictures of the rainbow, all the time shouting and bouncing around in the snow. "Holy cow, Colton, I see you! I see you."

He ran inside and fetched his cell phone and phoned Bill and Kaylee. "I'm looking at a rainbow over the rig where Colt fell," he said. "You wouldn't believe it. I've taken photos for you."

Bill said, "Ain't it a bit soon for Colton to be starting his angel act?"

Kaylee said, "Since when did Colton wait around for anything?"

And then the light shifted and all of a sudden the bright column of rainbow was gone, taking with it almost all of that rough Wyoming magic. And then, after that and for days and days and months and months to follow, there were only ordinary days—commonplace Wyoming blowing-empty days.

49

A Million-dollar Personality

Here's Colton's coffin being carried up the snowy path of the Evanston LDS South Stake Center so pitiful slow it takes your heart with it to watch the men, as if the weight on their shoulders is more than mortal man can ever carry in all this pressing February sunlight. You will recognize, by their shoulders alone, the men who act as pall-bearers from this story. There's Preston and he's taking it on the chin, as always, so his shoulders are square and solid. And you can see Bill who has folded up his feelings so they are not available to the reading public, but his heart is just about to burst, carrying his boy out in public like this, in a box. There's Jake, a good half a foot shorter than the tall Bryant men. The coffin barely weighs on him, but his soul feels tied to the earth, it's so heavy. And Tony is there, wishing he could carry Tabby's hurt along with the coffin, and JR, Colton's school friend, is also under the shiny, sharp corners of wood, trying to keep the thing moving forward and almost knee-buckled at the thought of how quickly they all went from boys to this.

People stop by Kaylee on their way into the church. "A million-dollar personality," says one.

"A great guy," says another.

"He was one in a million."

Kaylee smiles and nods but she doesn't say anything. She's afraid if she opens her mouth she will start screaming and not be able to stop. Jake has given her a bottle of Tylenol PM. "Colton gave me one of these after my car crash," he said. "You should try one. They'll knock you out."

Kaylee said, "I don't believe in drugs."

"Me either," said Jake, "but I think Colton would want you to take these for a few nights."

Now Kaylee is wondering if it's the Tylenol that's making her feel untethered. Someone puts a hand on her arm and says something and she nods. She feels as if maybe she could stop breathing and float skyward and be with Colton.

"Are you okay?" Someone else presses her arm.

And Kaylee doesn't want to let Colton down, not for anything. "Sure," she says. "Cowboy up, cupcake," she says.

Melissa sits in the front of the church, pale and tiny, watching the place where they will bring the coffin. She's trying not to think about where she is, trying not to feel this. If she doesn't feel it, she reasons, maybe it isn't happening. But then there's the coffin and the pallbearers in their pressed plaid shirts and blue jeans and Bill's wearing the custom-made cowboy boots Colton bought him for Christmas. The world is getting a lot blurry for Melissa and her heart feels as if it's being torn through her throat.

Jake stands to give the family prayer and then they play the song "Fishing in the Dark" and well over half the congregation starts to cry and most of the men don't even seem to care who knows it. The church is full to bursting. They've had to open up the gym too and that's also full to bursting. They're all there, every last single one of them. The little Kmart Cowboys, all grown up and looking at their shoes right about now. And there's those girls from school. There's most of Colton's crew from his days of flow testing and almost everyone from the rig. There are his teachers and ex-bosses and the parents of all his friends.

Merinda, Tabby, and Preston give their dedications and all the time Dakota doing somersaults and cartwheels at their feet and then Jake stands up to give the formal remarks and he tells all the

Colton stories. He tells about how Colton was teased as a kid and came up with the magical words, "Mind over matter," to deal with the pain of almost anything that happened to him. Jake talks about Cocoa and about how she ran away. He tells about the time Colton tried to retrieve a goose for him and he tells about Colton stopping the train. He has considered telling the story about how Colton had his Old Glory hanging out that day on the cliff, but he decides against it. Although he does tell how Colton was the most unlikely saint that God could have sent to earth. And now the *whole* congregation has dissolved in tears and men are openly weeping and not even bothering to wipe their tears with their sleeves and the presiding bishop, who truly didn't know Colton from a badger hole, looks out at the sea of stricken faces and his heart contracts so that when he stands up he finds himself talking about how Colton was sent down to teach all of us about forgiveness.

And then they play Sara Evans's "No Place That Far" and everyone piles back out into the fierce winter sunlight.

The family and a few friends drive over to the cemetery and Jake does the graveside dedication and they bury Colton in the trousers Melissa's mother just gave him for Christmas a month and a half ago. They have filled his pockets with ketchup packets and toy F350 Powerstroke pickups and they have stuffed a toy elk and a toy horse into the coffin with him. And then it's over.

Or it's just beginning.

Either way, Colton's gone.

50

EVANSTON CEMETERY

Evanston, Wyoming

It's been well over a year since Colton was killed, but you can still easily spot his site from anywhere else, spilling over with offerings like it had been freshly laid and still more freshly grieved upon. It's as if Colton and his grave stand in for every boy who died too young from the violence of Wyoming's way of life—cans of chewing tobacco, gallons of Mountain Dew, everything you can think of to do with hunting, horses, fishing, mountains, guns, and trucks. There is a hanging basket on a trellis next to flowers and a black ball cap (bleached by the sun to the color of sucked licorice in places) with the words western petroleum embroidered onto its front in orange thread. A plaster Labrador retriever sits patiently at the head of the stone with a solar lamp in its mouth, so that the grave will always have light.

A little after lunch on this day in early summer, Melissa comes with the boys. She has brought more bottles of Mountain Dew. Now they stretch out in a riot of green and yellow for half a foot on either side of the stone and threaten to take over the surrounding graves. She adds a few little packets of ketchup to the collection of condiments. Melissa gives the plaster dog a pat on the head. "You keeping an eye on him?" she asks. Melissa kneels in the grass next to the dog and watches Dakota and Nathanial run in and out and in and out of the trees. "Look," says Melissa. She holds up her wrist to the stone. "Like my tattoo?" She has had a bracelet inscribed in blue ink that reads in linking, fancy letters, "Mind over matter." She smiles through her tears. "Well dang it all, Colton H. Bryant, I surely like it." Then she wipes her eyes and says, "Holy cow, Colt . . . I sure miss you."

Someone is trotting a horse down the street behind the cemetery gates. It's calling out to a small herd in the scruffy paddock near the junction.

"Cowboy up, cupcake, I guess," says Melissa.

Birds flutter in and out of the trees and hop about on the lawn.

"I'll be getting on, then," says Melissa.

She gets to her feet and calls the boys. "Time to go home," she says. She buckles Dakota into his car seat. "Who's a big cowboy?" she asks him.

"Doggy," says Dakota, pointing to Colton's grave.

"Yup," says Melissa.

"Doggy!" Dakota screams.

"I know," says Melissa.

. . .

Later, in the early evening, Merinda and Shad, Tabby and Tony drive over to see Colton's grave.

"I don't even know who left half this stuff," says Tabby.

"Who brought the Copenhagen?" says Merinda, picking up a can of chew that has been propped up next to several toy trucks, a tiny horseshoe, tickets to a football game. "Did he even like the Cowboys that much?" she asks.

Tabby frowns. "People are kinda making a mess of Colton's grave. I'm going to clean some of this stuff up."

"You can't do that," says Merinda. "Then they will come back and their feelings will be hurt if their stuff is gone."

Tabby sighs. "I guess." Then she gets on her knees and clears away an arrangement of flowers from the stone so that the words on it can be seen. It's a slab of flawless Dakota stone set level to the earth engraved with mountains and an elk and a fir tree, a fish leaping from a river meandering through the bottom of the picture. It's exactly the country that Colton liked to cover with Cocoa on his weeks off the rigs. "Father, Husband, Brother, Son," it says below his name, "Colton H. Bryant," and his dates, "June 10, 1980–Feb 15, 2006." And then, under that, "Mind Over Matter," which is what everyone thought should be on the stone, and "Love Ya Always," which is what Kaylee wanted, and then finally the words "Sons: Nate & Dakota."

Tabby stands up and presses her hands into the small of her back. She says, "If it's another boy, I'll name him Colton H. Bryant Ruiz."

Merinda says, "What?"

"How does that sound?"

"Tabby!" says Merinda. "You're not . . .?"

Tabby says, "If it's a girl, it'll have to have Colt as a middle name . . ."

Merinda puts her hand over her mouth and looks from Tabby to Tony and back to Tabby again.

Tabby's smiling and looking down at Colton's grave.

Merinda says, "You're . . .?"

"Yup."

"Oh my glory," says Merinda. She bounces up and down a few times. "Holy cow, Tabby! You superstar! When is it due?"

Tony looks at his watch. "In about seven months, three weeks, four hours, and fifteen minutes," he says.

Tabby pushes him in the shoulder.

Merinda covers her ears and bounces a few more steps in front of Colton's grave. "Whoa," she says. "I didn't need to hear details. No details please."

51

COLT

———◆———

He was born, going seventy miles an hour on Highway 6, near Payson, Utah, in the early hours of June 10, 1980, because, as Kaylee says, "There weren't any decent hospitals in southwest Wyoming in those days, not that it would have mattered as it turns out, given he was born on the front seat of a 1976 Ford Thunderbird." The birth certificate gives a mile marker as the place of birth.

Bill glances over to see the infant on Kaylee's shoulder, already lifting its head, staring up at the streetlights. "Cover its eyes. I don't think they're supposed to look at such bright lights so soon."

Kaylee puts her hand over the back of the baby's head, steadying him, and makes it so that he can't look up. "Whew," she says. Fifteen minutes, if that, between the first sign he was on his way and now here he is and everything he isn't pooling darkly on the passenger seat under Kaylee's legs. "So help me, whatever I'm sitting in."

Bill gives Kaylee one of his specialty sideways smiles. "Well, what we got now?" he asks.

"I can't tell. Hold on." Kaylee lifts the baby up to a passing streetlight. "Boy," she announces.

"Then there you go. There's your colt you been pestering me for."

Kaylee's laughing. "So let's call him Colton," she says. Colton pulls away and begins to paddle, as if trying to feel the limits of his new world and, finding none, trying to swim away on his own umbilical pull back to the earth. Kaylee tries on the name, "Colton Bryant."

"Colton *H.* Bryant," says Bill.

There's a pause. "What's the 'H.' for?" asks Kaylee.

"Show," says Bill. "Just for show."

52

JAKE AND COLTON

Afterwards

⸺⸺⋅⊱⋆⊰⋅⸺⸺

A month after it would have been Colton's twenty-seventh birthday, a little less than a year and a half after he died, Jake says to Tonya, "I think I'll just head out for some air." They've put the kids to bed and cleaned up the kitchen and they've spent a couple of hours working on the new porch. They are getting ready for bed themselves.

Tonya nods. "Okay."

Jake pulls a ball cap onto his head and takes his keys off the counter.

"Say hi to him for me."

"Yep."

Jake gets in his truck, a white F250, and drives a mile or two southeast onto Highway 191, then he turns right onto Paradise Road just this side of the New Fork River. There is a whole cluster of rig signs swinging in the wind, including the circular orange, black, and white sign that reads patterson-uti drilling company—snyder, texas. On the tail of an arrow cut through the sign is the number 455. That was Colton's rig, moved on to a fresh location. Another sign hangs close to this one except this has the numbers 515 on the tail of its arrow. That's rig 455's sister rig. Jake doesn't even pay close attention to the signs. He's done this drive a lot since Colton died. He knows his way around this oil patch blindfolded.

It's dark by now. Jake rolls down the window and taps his fingers on the steering wheel with the radio tuned to the country station. Anchored to the shadowy swell of the high plains there are maybe fifteen, twenty drilling rigs, lit up like so many Eiffel Towers, with fresh-cut roads like veins going deep into the high plains to the heart of each pad. The headlights of so many company trucks bob along in the darkness, like lost, disembodied orbs looking for a place to roost. Jake drives a distance toward the rigs on a freshly paved road and when he hits the dirt he turns right onto another dirt road, past a sign that says nerd farms and another sign that says the land beyond this point is Bureau of Land Management critical winter habitat and that traffic is prohibited here from December until April, except for activity related to drilling. The air here is sour with gas. Jake keeps driving.

The radio gets jumpy with static. Jake switches it off and now he can hear the strange singing of the drilling rigs on either side of the road—a high, breathy sound—as if the drills might be calling softly to one another, a fresh breed of metal animal, caroling the way wolves and coyotes used to do out here. Now the smell of gas is overwhelming, like walking into a chemistry lab of broken Bunsen burners. Fans in a nearby compressor station scream. And then, faint and lonely through the midsummer night sky, there is the occasional shout of one roughneck to another.

They say Indians used to press their ears to the earth to hear what was coming next, but if you were to lie down and listen to this ground, you'd get run over. Jake turns right at the compressor station and now he's facing the Wind River Mountains again. A hailstorm this afternoon pulverized the drought-stricken sage and a cloud of it hangs about a foot above the ground in a grey-green mist. Jake turns right again. You wouldn't know, if you didn't know where to look, where Colton fell. There's nothing to mark the spot except an abandoned well, covered over, taped up. A sign saying riverside 88-2d hangs above one of the pipes. Jackrabbits and a couple of pronghorn antelope scuffle about on the surrounding gravel.

Strands of rope hung with yellow, red, and blue pennants—like triangular prayer flags without the prayers—flap over the abandoned evaporating pond. Jake gets out of his truck and faces the sky to the west. He takes a deep breath of the gas-smelling air and puts his hands together. "Hey Colt," he says.

Nothing much happens. The jackrabbits hop off a few more steps and the pronghorns jolt a little further away and then look over their shoulders at Jake.

Jake bows his head and puts his hands over his eyes. "I know you're in a better place," he says. "So I'm not so sad. Kinda."

Above Jake's shoulder, to the northeast, the two Patterson-UTI rigs, 455 and 515—red, white, and blue in the daylight, but anonymously lit up at night, just like all the other rigs—keep on drilling and drilling into the earth. Fifteen thousand feet deep and beyond for natural gas that will go to the compressor stations less than a quarter of a mile from where Colton fell and from there into pipes that run, but not fast enough to keep up with demand, across to the coast to fulfill California's demand for the promised, big, hot-and-cold on-demand life.

The last glow of the sun shuts down but the sky remains an anonymous blur of grey. There aren't shooting stars above the plains anymore. Now the brightest places in the plains are the rigs, all violent with night lights. The antelope disappear into shadows.

Jake says, "Tonya says hi. Me and her are building on a porch out back." He smiles and says, "But I guess you can see all that from up there, huh?" Jake kicks the dirt with the toe of his boot and spits. "Yup," he says, "I ain't doing so bad. Kids are doing good. We're thinking maybe of getting another one. Huh, what you think about that?" Then Jake clears his throat and wipes his nose on his sleeve. "Well," he says, "that about does 'er, I guess." He climbs back into his pickup truck and drives back the way he came. "Yep," he says, "that's about all there is to say," and he's glad it's dark and that he's alone in the truck because some people don't like to see a grown man cry.

Author's Note

This is a work of nonfiction, but I have taken narrative liberties with the text. I have emphasized certain aspects of Colton's life and of his personality and disregarded others. I have re-created dialogue and occasionally juggled time to create a smoother story line. I have changed one name (Chase). I must emphasize, however, that Colton's friends and family were never less than honest and open with me and they were endlessly patient and understanding with my questions, some of which can only have been incredibly painful.

Colton was the fourth rig hand in just over eighteen months in the Upper Green River Valley to die at an Ultra Petroleum well site. Ultra Petroleum has made it clear, through their website and other announcements, that profit is their priority. They have repeatedly boasted that they have the lowest cost per thousand cubic feet of gas produced in the industry. In July 2005, Brian Ault, the vice president of Ultra Petroleum, quit the company. "It makes me sick," he was reported as saying, giving his reasons for leaving UP, "how much we're pulling out of the ground and how little we're giving back."

The week before Colton died, Michael D. Wattford, chairman, president, and chief executive officer of Ultra Petroleum, issued a statement that reads, in part, "Ultra Petroleum continues to excel. 2005 was another record year in a continuing string of record reporting periods. . . . Our 2006 plans continue the growth theme. . . . We plan to execute the most aggressive drilling program in our history with 160 gross wells in Wyoming, and bring into production three more fields in China. With over seventeen years of identifying drilling opportunities in Wyoming *coupled with our low cost structure,* we remain positioned to continue delivering industry leading performances for many years to come" (emphasis added).

For their part in Colton Bryant's accident, in which six serious safety violations were found, Wyoming Occupational Health and Safety (OSHA) fined Patterson-UTI $7,031. On May 3, 2006, less than three months after Colton's death, Patterson-UTI Energy announced record results for the first quarter of 2006. Their net income for the quarter had increased 174 percent to $159 million and their revenues for the quarter were up by 70 percent to $598 million. Ultra Petroleum was not found culpable of any infractions in relation to the accident that killed Colton. In 2006, revenues at Ultra hit a record $592.7 million. Beyond worker's compensation, Colton's family received no compensation for his loss.

In his article "Fatalities in the energy fields: 2000–2006" (*High Country News,* vol. 39, no. 6, April 2, 2007), Ray Ring writes, "At least 89 people died on the job in the Interior West's oil and gas industry from 2000 to 2006 in a variety of accidents, including 90-foot falls, massive explosions, poison gas inhalations and crushings by safety harnesses." Of those deaths, Wyoming is responsible for thirty-five, by far the highest percentage.

In April 2007, the *Casper Star-Tribune* reported that Wyoming and Montana had the worst records in the nation for workplace safety in 2005. "Wyoming has the highest rate of job fatalities with 16.8 per 100,000 workers and Montana has the second highest, with 10.3 fatalities per 100,000 workers." Since 2006, the State of Wyoming Department of Employment has increased its personnel to eight OSHA compliance officers working in the

entire state. Four are based in Cheyenne and four in the field. No
rig hand I spoke to beyond those associated with a fatality on the
rigs—even those who had worked in the industry for decades—
has ever seen an OSHA compliance officer on the oil patch. It
would have cost Patterson-UTI two thousand dollars, at most, to
have safety rails where Colton fell.

ACKNOWLEDGMENTS

I can never adequately express my gratitude to Colton's family for trusting me with his story; Bill and Kaylee Bryant, Tabby and Tony Ruiz, Merinda Bryant and Shad Powers, Melissa Bryant, Preston and Mandi Bryant. For incredible generosity and kindness, my thanks to Jake and Tonya Wigginton.

For all those who took me on cattle drives and onto the oil patch, and who explained their love of Wyoming to me, my deepest thanks: John and Lucy Fandek, Freddy Botur, Holly Davis, Saul Bencomo, Aaron "James Curry," Linda Baker, Steve Belinda, Bill Close, John Carney, Linda Goodman, Joel and Kim Berger, and Bridget Mackey. For socioeconomic statistics and endless patience with my questions, Jeffrey Jacquet.

For reading parts and/or the whole of this manuscript in all its incarnations and for encouragement, support and guidance during the writing of it: Joan Blatt, Bryan Christy, Nicola Fuller, Terry Tempest Williams, Oliver Payne, Dean Stayner, David Baron, Adanna Moriarty, Dan Glick, Bill Broyles, Cassie and Bill Ross, and James Galvin.

For their advice and guidance and faith, thanks to Ann Godoff and Melanie Jackson.

And for their love and patience and support, deepest thanks to my husband, Charlie Ross, and to my children Sarah, Fuller, and Cecily Ross.

ABOUT THE AUTHOR

Alexandra Fuller was born in England in 1969. In 1972 she moved with her family to a farm in southern Africa where she lived until her mid-twenties. She has lived in Wyoming with her husband since 1994. They have three children.